R for
Statistics

R for Statistics

Pierre-André Cornillon

Arnaud Guyader

François Husson

Nicolas Jégou

Julie Josse

Maela Kloareg

Eric Matzner-Løber

Laurent Rouvière

CRC Press
Taylor & Francis Group
Boca Raton London New York

CRC Press is an imprint of the
Taylor & Francis Group, an **informa** business

A CHAPMAN & HALL BOOK

CRC Press
Taylor & Francis Group
6000 Broken Sound Parkway NW, Suite 300
Boca Raton, FL 33487-2742

© 2012 by Taylor & Francis Group, LLC
CRC Press is an imprint of Taylor & Francis Group, an Informa business

No claim to original U.S. Government works

Printed in the United States of America on acid-free paper
Version Date: 20120113

International Standard Book Number: 978-1-4398-8145-3 (Paperback)

Visit the Taylor & Francis Web site at
http://www.taylorandfrancis.com

and the CRC Press Web site at
http://www.crcpress.com

Contents

Preface   xiii

I   An Overview of R   1

1   Main Concepts   3

1.1	Installing R	3
1.2	Work Session	3
	1.2.1 Session in Linux	4
	1.2.2 Session in Windows	4
	1.2.3 Session on a Mac	5
1.3	Help	5
	1.3.1 Online Assistance	5
	1.3.2 Help on CRAN	6
1.4	R Objects	6
	1.4.1 Creating, Displaying and Deleting Objects	6
	1.4.2 Type of Object	7
	1.4.3 The Missing Value	8
	1.4.4 Vectors	9
	1.4.4.1 Numeric Vectors	9
	1.4.4.2 Character Vectors	10
	1.4.4.3 Logical Vectors	11
	1.4.4.4 Selecting Part of a Vector	12
	1.4.4.5 Selection in Practice	13
	1.4.5 Matrices	13
	1.4.5.1 Selecting Elements or Part of a Matrix	14
	1.4.5.2 Calculating with the Matrices	16
	1.4.5.3 Row and Column Operations	17
	1.4.6 Factors	18
	1.4.7 Lists	20
	1.4.7.1 Creating a List	20
	1.4.7.2 Extraction	20
	1.4.7.3 A Few List Functions	21
	1.4.7.4 Dimnames List	21
	1.4.8 Data-Frames	22
1.5	Functions	23
	1.5.1 Arguments of a Function	23

v

      1.5.2   Output 24
  1.6 Packages 24
      1.6.1   What Is a Package ? 24
      1.6.2   Installing a Package 24
      1.6.3   Updating Packages 25
      1.6.4   Using Packages 25
  1.7 Exercises 26

2 Preparing Data   29
  2.1 Reading Data from File 29
  2.2 Exporting Results 33
  2.3 Manipulating Variables 34
      2.3.1   Changing Type 34
      2.3.2   Dividing into Classes 35
      2.3.3   Working on Factor Levels 36
  2.4 Manipulating Individuals 39
      2.4.1   Identifying Missing Data 39
      2.4.2   Finding Outliers 42
  2.5 Concatenating Data Tables 43
  2.6 Cross-Tabulation 46
  2.7 Exercises 48

3 R Graphics   51
  3.1 Conventional Graphical Functions 51
      3.1.1   The Plot Function 52
      3.1.2   Representing a Distribution 58
      3.1.3   Adding to Graphs 60
      3.1.4   Graphs with Multiple Dimensions 64
      3.1.5   Exporting Graphs 66
      3.1.6   Multiple Graphs 67
      3.1.7   Multiple Windows 69
      3.1.8   Improving and Personalising Graphs 70
  3.2 Graphical Functions with lattice 73
      3.2.1   Characteristics of a "Lattice" Graph 75
      3.2.2   Formulae and Groups 76
      3.2.3   Customising Graphs 79
            3.2.3.1   Panel Function 79
            3.2.3.2   Controlling Size 80
            3.2.3.3   Panel Layout 81
      3.2.4   Exportation 81
      3.2.5   Other Packages 82
  3.3 Exercises 82

4 Making Programs with R   **87**

4.1   Control Flows 87
     4.1.1   Grouped Expressions 87
     4.1.2   Loops (for or while) 87
     4.1.3   Conditions (if, else) 89

4.2   Predefined Functions 90

4.3   Creating a Function 97

4.4   Exercises 100

II   Statistical Methods   **101**

Introduction to the Statistical Methods Part   **103**

5 A Quick Start with R   **105**

5.1   Installing R 105

5.2   Opening and Closing R 105

5.3   The Command Prompt 105

5.4   Attribution, Objects, and Function 106

5.5   Selection 106

5.6   Other 107

5.7   Rcmdr Package 107

5.8   Importing (or Inputting) Data 107

5.9   Graphs 108

5.10   Statistical Analysis 108

6 Hypothesis Test   **109**

6.1   Confidence Intervals for a Mean 109
     6.1.1   Objective 109
     6.1.2   Example 109
     6.1.3   Steps 109
     6.1.4   Processing the Example 110
     6.1.5   Rcmdr Corner 112
     6.1.6   Taking Things Further 112

6.2   Chi-Square Test of Independence 113
     6.2.1   Objective 113
     6.2.2   Example 113
     6.2.3   Steps 113
     6.2.4   Processing the Example 114
     6.2.5   Rcmdr Corner 116
     6.2.6   Taking Things Further 117

6.3   Comparison of Two Means 118
     6.3.1   Objective 118
     6.3.2   Example 118
     6.3.3   Steps 119
     6.3.4   Processing the Example 119
     6.3.5   Rcmdr Corner 123

	6.3.6	Taking Things Further	124
6.4	Testing Conformity of a Proportion		125
	6.4.1	Objective	125
	6.4.2	Example	125
	6.4.3	Step	125
	6.4.4	Processing the Example	125
	6.4.5	Rcmdr Corner	126
	6.4.6	Taking Things Further	126
6.5	Comparing Several Proportions		127
	6.5.1	Objective	127
	6.5.2	Example	127
	6.5.3	Step	127
	6.5.4	Processing the Example	127
	6.5.5	Rcmdr Corner	128
	6.5.6	Taking Things Further	128
6.6	The Power of a Test		129
	6.6.1	Objective	129
	6.6.2	Example	129
	6.6.3	Steps	129
	6.6.4	Processing the Example	130
	6.6.5	Rcmdr Corner	130
	6.6.6	Taking Things Further	131

7 Regression   **133**
7.1	Simple Linear Regression		133
	7.1.1	Objective	133
	7.1.2	Example	134
	7.1.3	Steps	134
	7.1.4	Processing the Example	134
	7.1.5	Rcmdr Corner	138
	7.1.6	Taking Things Further	139
7.2	Multiple Linear Regression		140
	7.2.1	Objective	140
	7.2.2	Example	141
	7.2.3	Steps	141
	7.2.4	Processing the Example	141
	7.2.5	Rcmdr Corner	145
	7.2.6	Taking Things Further	146
7.3	Partial Least Squares (PLS) Regression		147
	7.3.1	Objective	147
	7.3.2	Example	147
	7.3.3	Steps	148
	7.3.4	Processing the Example	148
	7.3.5	Rcmdr Corner	153
	7.3.6	Taking Things Further	153

8 Analysis of Variance and Covariance   **157**
  8.1 One-Way Analysis of Variance   157
       8.1.1 Objectives   157
       8.1.2 Example   158
       8.1.3 Steps   158
       8.1.4 Processing the Example   159
       8.1.5 Rcmdr Corner   164
       8.1.6 Taking Things Further   165
  8.2 Multi-Way Analysis of Variance with Interaction   166
       8.2.1 Objective   166
       8.2.2 Example   167
       8.2.3 Steps   167
       8.2.4 Processing the Example   168
       8.2.5 Rcmdr Corner   171
       8.2.6 Taking Things Further   172
  8.3 Analysis of Covariance   173
       8.3.1 Objective   173
       8.3.2 Example   174
       8.3.3 Steps   175
       8.3.4 Processing the Example   175
       8.3.5 Rcmdr Corner   179
       8.3.6 Taking Things Further   179

9 Classification   **181**
  9.1 Linear Discriminant Analysis   181
       9.1.1 Objective   181
       9.1.2 Example   182
       9.1.3 Steps   183
       9.1.4 Processing the Example   183
       9.1.5 Rcmdr Corner   188
       9.1.6 Taking Things Further   188
  9.2 Logistic Regression   190
       9.2.1 Objective   190
       9.2.2 Example   190
       9.2.3 Steps   191
       9.2.4 Processing the Example   191
       9.2.5 Rcmdr Corner   197
       9.2.6 Taking Things Further   198
  9.3 Decision Tree   199
       9.3.1 Objective   199
       9.3.2 Example   199
       9.3.3 Steps   199
       9.3.4 Processing the Example   199
       9.3.5 Rcmdr Corner   208
       9.3.6 Taking Things Further   208

10 Exploratory Multivariate Analysis   **209**
   10.1 Principal Component Analysis   209
        10.1.1 Objective   209
        10.1.2 Example   209
        10.1.3 Steps   210
        10.1.4 Processing the Example   210
        10.1.5 Rcmdr Corner   218
        10.1.6 Taking Things Further   219
   10.2 Correspondence Analysis   222
        10.2.1 Objective   222
        10.2.2 Example   222
        10.2.3 Steps   222
        10.2.4 Processing the Example   223
        10.2.5 Rcmdr Corner   227
        10.2.6 Taking Things Further   228
   10.3 Multiple Correspondence Analysis   230
        10.3.1 Objective   230
        10.3.2 Example   230
        10.3.3 Steps   231
        10.3.4 Processing the Example   231
        10.3.5 Rcmdr Corner   239
        10.3.6 Taking Things Further   239

11 Clustering   **241**
   11.1 Ascending Hierarchical Clustering   241
        11.1.1 Objective   241
        11.1.2 Example   241
        11.1.3 Steps   241
        11.1.4 Processing the Example   242
        11.1.5 Rcmdr Corner   246
        11.1.6 Taking Things Further   247
   11.2 The k-Means Method   252
        11.2.1 Objective   252
        11.2.2 Example   252
        11.2.3 Steps   252
        11.2.4 Processing the Example   252
        11.2.5 Rcmdr Corner   255
        11.2.6 Taking Things Further   256

Appendix   **257**
   A.1 The Most Useful Functions   257
        A.1.1 Generic Functions   257
        A.1.2 Numerical Functions   257
        A.1.3 Data Handling Functions   258
        A.1.4 Probability Distributions   260

A.1.5 Basic Statistical Functions 261
A.1.6 Advanced Statistical Functions 262
A.1.7 Graphical Functions 263
A.1.8 Import and Export Functions 264
A.1.9 Text Management 265
A.1.10 Some Other Useful Functions 265
A.2 Writing a Formula for the Models 266
A.3 The Rcmdr Package 267
A.4 The FactoMineR Package 269
A.4.1 The FactoMineR Package 269
A.4.2 The Drop-Down Menu 270
A.5 Answers to the Exercises 270
A.5.1 Exercises: Chapter 1 270
A.5.2 Exercises: Chapter 2 278
A.5.3 Exercises: Chapter 3 287
A.5.4 Exercises: Chapter 4 292

Bibliography **295**

Index of the Functions **297**

Index **301**

Preface

This book is the English adaptation of the second edition of the book *Statistiques avec R* which was published in 2008 and was a great success in the French-speaking world. In this version, a number of worked examples have been supplemented and new examples have been added. We hope that readers will enjoy using this book for reference when working with R.

This book is aimed at statisticians in the widest sense, that is to say, all those working with datasets: science students, biologists, economists, etc. All statistical studies depend on vast quantities of information, and computerised tools are therefore becoming more and more essential. There are currently a wide variety of software packages which meet these requirements. Here we have opted for R, which has the triple advantage of being free, comprehensive, and its use is booming. However, no prior experience of the software is required. This work aims to be accessible and useful for both novices and experts alike.

This book is organised into two main parts: Part I focuses on the R software and the way it works, and Part II on the implementation of traditional statistical methods with R. In order to render them as independent as possible, a brief chapter offers extra help getting started (Chapter 5, "A Quick Start with R") and acts as a transition: It will help those readers who are more interested in statistics than in software to be operational more quickly.

In the first chapter we present the basic elements of the software. The second chapter deals with data processing (reading data from file, merging factor levels, etc.), that is to say, common operations in statistics. As any statistical report depends on clear, concise visualisation of results, in Chapter 3 we detail the variety of possibilities available in this domain with R. We first present the construction of simple graphs with the various available options, and then detail the use of more complex graphs. Some programming basics are then presented in Chapter 4: We explain how to construct your own functions but also present some of the useful pre-defined functions to conduct repetitive analyses automatically. Focusing on the R software itself, this first main part enables the reader to understand the commands used in subsequent statistical studies.

The second part of the book offers solutions for dealing with a wide range of traditional statistical data processing techniques: hypothesis tests (Chapter 6), regression methods (Chapter 7), analyses of variance and co-variance (Chapter 8), classification methods (Chapter 9), exploratory multi-variate analysis methods (Chapter 10), and clustering methods (Chapter 11). Each technique is preceded by a concrete example and is dealt with independently in its own specific worked example. Following a short presentation of the method, the R command lines are explicitly detailed and the results commented. Readers are able to download the datasets from the address http://www.agrocampus-ouest.fr/math/RforStat, where they can easily find all the results described in this work. R is a command line interface, but another easy-to-use graphical interface is available: At the end of each worked example there is a section called "Rcmdr Corner" which can be used to process the example using this second interface. This section also helps readers get to know the R commands better as, with each click on the drop-down menu, the lines of code corresponding to the chosen action are generated.

To draw this preface to a close, we would like to thank Jérôme Pagès and Dominique Dehay, who enabled us to write the French version under the best possible conditions. We would also like to thank Rebecca Clayton for the English translation and Agrocampus Ouest for financing this translation.

Part I

An Overview of **R**

1

Main Concepts

This chapter provides a general presentation of R software. We first describe how to install it, the differences in use depending on the operating system, and the different help available. We then go on to list the various elements used in R (vectors, matrices, etc.), as well as a few comments about their functions. We conclude by presenting the packages, or libraries, of external programs which will be useful throughout this book.

1.1 Installing **R**

R is an open source software environment for statistics freely distributed by CRAN (Comprehensive R Archive Network) at the following address http://cran.r-project.org/. The installation of R varies according to the operating system (Windows, Mac OS X, or Linux) but the functions are exactly the same and most of the programs are portable from one system to another. Installing R is very simple, just follow the instructions.

1.2 Work Session

To use R, first open a work session. R then opens a window with the following notice:

```
R version 2.14.1 (2011-12-22)
Copyright (C) 2011 The R Foundation for Statistical Computing
ISBN 3-900051-07-0
Platform: i386-pc-mingw32/i386 (32-bit)

R is free software and comes with ABSOLUTELY NO WARRANTY.
You are welcome to redistribute it under certain conditions.
Type 'license()' or 'licence()' for distribution details.

  Natural language support but running in an English locale
```

```
R is a collaborative project with many contributors.
Type 'contributors()' for more information and
'citation()' on how to cite R or R packages in publications.

Type 'demo()' for some demos, 'help()' for on-line help, or
'help.start()' for an HTML browser interface to help.
Type 'q()' to quit R.

>
```

R then waits for an instruction, as indicated by the symbol ">" at the beginning of the line. Each instruction must be validated by Enter to be run. If the instruction is correct, R again waits for another instruction, as indicated by ">". If the instruction is incomplete, R yields the symbol "+". You must then complete the instruction or take control again by typing Ctrl + c or Esc. If the instruction is incorrect, an error message will appear.

For each project, we recommend you create a file in which an image of the session concerning this project will be saved. We also recommend that users write the commands in a text file in order to be able to use them again when needed.

By default, R saves all the objects created (variables, results tables, etc.). At the end of the session, these objects can be saved in an image of the session **.RData**, using the command **save.image()** or when closing a session. To close a session in R, use the **quit** function:

```
> q()
Save workspace image? [y/n/c]:
```

The backup file can be found in the project file (we will see later how to change the destination). Saved objects are thus available for future use using the **load** function. We will now look at the differences between a session opened in Linux, in Windows, and on a Mac.

1.2.1 Session in Linux

To open an R session, type the command R in a command window and indicate the path to the working directory using the command **setwd()**. We can also open an R session directly in the directory in which we wish to work. The destination can be checked using **getwd()**.

1.2.2 Session in Windows

To begin an R session, click on the R icon on the desktop to open a window. The data and the commands will be saved in the same place as R was opened, that is to say, in the same place as R is installed.

The best way to change the working directory is to use the **setwd()** function or by going to `File` then `Change dir....`

To close a session in R, go to `File`, select `Exit` (or equivalently type **q()** in the command window) and answer yes when asked `Save workspace image?` An R icon will then be created in the working directory. When you click on this icon, a new session will open directly in the given directory and the objects created in the previous session will be available.

1.2.3   Session on a Mac

There are two ways of opening an R session. The first is to open the working directory and type the R command. The second option is to click on the R icon in the Applications menu and an R window will open. The working data and the commands used must therefore be saved in the same place as R was opened. It is also possible to choose the directory in which we wish to work using the option `Change Working Directory` of the `Misc` icon.

To close a session in R, go to `R`, then `Quit R` and choose whether or not to save the session. A **.RData** file is created in the working directory. However, it is not possible to view hidden files with *Finder* and it is therefore impossible to click on this file.

1.3   Help

1.3.1   Online Assistance

For help with a given function, for example the **mean** function, simply use

> **help(mean)**

or

> **?mean**

Help is displayed directly in the interface. It is also possible to display help in HTML format in a web browser. For this option, type

> **help.start()**

HTML help offers a search function as well as functions grouped under different headings. This is probably the most appropriate help option for a novice user.

Within the help section for each function you will find the list of arguments to be detailed, along with their default value where appropriate, the list of results returned by the function, and the name of the programmer. At the end of the help section, one or more examples of how the function is used are presented and can be copied and pasted into R in order to better understand its use.

1.3.2   Help on CRAN

The help provided by R is not always sufficient. Indeed, sometimes for a specific question, a user will not find an appropriate solution. For this type of question, there is a website where users can search for solutions in the archives of R users. This help can be accessed from an R command:

```
> RSiteSearch("mean weighted")
```

or from the website `http://finzi.psych.upenn.edu/search.html`. A FAQ section is also available at the address `http://cran.r-project.org/doc/FAQ/R-FAQ.html`. Usually, the question has already been asked by other users and the answer to your problem can be found here. Finally, you can also search directly on the Internet using a search engine such as Google, with the keywords R or CRAN or use the website `http://www.rseek.org/` which is dedicated to R search on the Internet.

1.4   R Objects

R uses functions or operators which act on objects (vectors, matrices, etc.). All the functions in this book are written in bold. This section deals with presenting and manipulating different R objects.

1.4.1   Creating, Displaying and Deleting Objects

An object can be created by being assigned to one of the three operators "<-", "->", "=" and by naming the object:

```
> b<-41.3   # set up object b consisting of one number: 41.3
> x<-b      # b is assigned to x
> x=b       # b is assigned to x
> b->x      # b is assigned to x
```

For the rest of this book we use the assigning operator "<-". The symbol "#" indicates that all the rest of the line will not be interpreted by R; this is so that comments can be added to a program.

Existing R data can also be retrieved using the following function **data**:

> **data**(iris)

This command is used to load the `iris` dataset, which is now available. Data can also be read directly from a file using the function **read.table** (see Section 2.1), for example,

> **ozone** <- **read.table**("ozone.txt",header=TRUE)

If an object does not exist, the allocation creates it. If not, the allocation replaces the previous value without warning. We display the value of an object x via the command

> **print**(x)

or, more simply

> x

By default, R saves all the objects created in the session. We therefore recommend that you delete these objects regularly. To find the objects from a session, use the function **objects**() or **ls**(). To delete the object x, type

> **rm**(x)

and to delete multiple objects

> **rm**(object1,object2,object3)

Finally, to delete a list of objects for which a part of the name is common, for example the letter a, use

> **rm**(list=ls(pattern=".*a.*"))

1.4.2 Type of Object

Before looking at the different objects in R, it is important to know the main modes or types. These types are

- Null (empty object): `NULL`

- Logical (Boolean): `TRUE`, `FALSE` or `T`, `F`

- Numeric (real number): `1`, `2.3222`, `pi`, `1e-10`

- Complex (complex number): `2+0i`, `2i`

- Character (chain of characters): `'hi'`, `"K"`

In order to know the mode of an object x of R, simply run the command

> **mode**(x)

It is also possible to test which mode a particular object x belongs to. The results are Booleans with values that are either TRUE or FALSE:

> **is.null**(x)
> **is.logical**(x)
> **is.numeric**(x)
> **is.complex**(x)
> **is.character**(x)

It is possible to convert the object x of a given mode explicitly using the following commands:

> **as.logical**(x)
> **as.numeric**(x)
> **as.complex**(x)
> **as.character**(x)

However, we must be careful with regard to the meaning of these conversions. R always yields a result for every conversion instruction even if it is meaningless. For example, from the following conversion table:

From	To	Function	Conversions
Logical	Numeric	**as.numeric**	FALSE $\to$ 0
			TRUE $\to$ 1
Logical	Character	**as.character**	FALSE $\to$ "FALSE"
			TRUE $\to$ "TRUE"
Character	Numeric	**as.numeric**	"1", "2", ... $\to$ 1, 2, ...
			"A", ... $\to$ NA
Character	Logical	**as.logical**	"FALSE", "F" $\to$ FALSE
			"TRUE", "T" $\to$ TRUE
			Other characters $\to$ NA
Numeric	Logical	**as.logical**	0 $\to$ FALSE
			Other numerics $\to$ TRUE
Numeric	Character	**as.character**	1, 2, ... $\to$ "1", "2", ...

Each object has two intrinsic attributes: its mode **mode**() and its length **length**(). There are also other specific attributes which vary according to the type of object: **dim**, **dimnames**, **class**, **names**. It is possible to request a list of these specific attributes by running the command

> **attributes**(object)

1.4.3   The Missing Value

For a number of reasons, certain elements of data may not be collected during an experiment or study. These elements are known as missing data. They are therefore not available for the user, and R denotes them NA for "Not Available". This is not a true mode and it has its own calculation rules. For example,

```
> x <- NA
> print(x+1)
> NA
```

In order to know where to find the missing values for an object x, we must ask the question:

```
> is.na(x)
```

This yields a Boolean of the same length as x. The question is also asked element by element. In the case of a vector, this yields a logical vector the same length as x with TRUE if the corresponding element in x is NA, and FALSE if not.

Other special values should be noted here: Inf for infinity and NaN for "Not a Number", values resulting from calculation problems; for example **exp**(1e10) or **log**(-2), respectively.

1.4.4 Vectors

Vectors are atomic objects, that is to say, they are of a unique type (null, logical, numeric, etc.), made up of a set of values called components, coordinates or elements. The attribute length, obtained using the function **length**, gives the number of elements in the vector.

1.4.4.1 Numeric Vectors

There are a number of different ways of constructing a vector, some of which are detailed below:

- Construction by the collect function **c**:

```
> x <- c(5.6,-2,78,42.3) # numeric vector with 4 entries
> x
[1]   5.6 -2.0 78.0 42.3
> x <- c(x,3,c(12,8))     # vector with 7 entries
> x
[1]   5.6 -2.0 78.0 42.3  3.0 12.0  8.0
> x <- 2 # vector of length 1
> x
[1] 2
```

- Construction by sequence operator ":":

```
> 1:6
[1] 1 2 3 4 5 6
```

- Construction by the **seq** function:

```
> seq(1,6,by=0.5)
 [1] 1.0 1.5 2.0 2.5 3.0 3.5 4.0 4.5 5.0 5.5 6.0
> seq(1,6,length=5)
 [1] 1.00 2.25 3.50 4.75 6.00
```

- Construction by the **rep** function:

```
> rep(1,4)
 [1] 1 1 1 1
> rep(c(1,2),each=3)
 [1] 1 1 1 2 2 2
```

- Construction by the **scan** function. R will then ask you to enter the elements one at a time. Using the argument **n=3**, we specify that three elements will be collected. If we do not specify **n**, the collection is stopped by putting in an empty value.

```
> marks <- scan(n=3)
1: 12
2: 18
3: 15
Read 3 items
```

1.4.4.2   Character Vectors

It is possible to create character vectors in the same way with the functions **c** or **rep**. For example,

```
> x <- c("A","BB","C1")
> x
[1] "A"  "BB" "C1"
> x <- rep('A',5)
> x
[1] "A" "A" "A" "A" "A"
```

Even if R interprets " and ' in the same way, we will use " from here on in. It is also possible to create character vectors using the vector `letters` which contains the letters of the alphabet. The function **format** is used to organise numerical data in chains of characters of the same length (see also the **toString** function).

Another way of doing this is to manipulate different R objects and to concatenate them or to extract part of them. To concatenate the objects, use the **paste** function:

```
> paste("X",1:5,sep="-")
[1] "X-1" "X-2" "X-3" "X-4" "X-5"
```

```
> paste(c("X","Y"),1:5,"txt",sep=".")
[1] "X.1.txt"  "Y.2.txt"  "X.3.txt"  "Y.4.txt"  "X.5.txt"
> paste(c("X","Y"),1:5,sep=".",collapse="+")
[1] "X.1+Y.2+X.3+Y.4+X.5"
```

The argument `collapse` groups together all the elements in a vector with a length of 1. For an extraction, use the function **substr**:

```
> substr("freerider",5,9)
[1] "rider"
```

This extracts the characters ranked from 5 to 9 from `freerider` which yields `"rider"`.

1.4.4.3 Logical Vectors

Boolean vectors are usually generated with logical operators: ">", ">=", "<", "<=", "==", "!=", etc. They can also be generated using the functions **seq**, **rep** and **c**. They can be used to make complex selections (see Section 1.4.4.4, p. 12) or conditions operations (see Section 4.1, p. 87). Let us examine the following example:

```
> 1>0
[1] TRUE
```

This command yields a logical vector with a length of 1 which is `TRUE`, as 1 is greater than 0. The command

```
> x>13
```

yields a logical vector of the same length as `x`. Its elements have the value `TRUE` when the corresponding element meets the given condition (here strictly higher than 13) and the value `FALSE` if it does not. During the calculations, the logicals are changed into numerics, following the convention that the `FALSE` become 0 and the `TRUE` become 1. Let us examine an example of this. An object `test` is created, which is the vector of the logicals (`FALSE,FALSE,TRUE`). We then calculate the following product:

```
> x <- c(-1,0,2)
> test <- x>1
> (1+x^2)*(x>1)
[1] 0 0 5
```

The functions **all** and **any** can also be used. The function **all** yields `TRUE` if all the elements meet the condition and if not, it yields `FALSE`. The function **any** yields `TRUE` if one of the elements meets the condition, and if not, `FALSE`:

```
> all(x>1)
[1] FALSE
> any(x>1)
[1] TRUE
```

1.4.4.4 Selecting Part of a Vector

Selections are made using the selection operator [ ] and a selection vector:

```
> x[indexvector]
```

The selection vector can be a vector of positive integers, negative integers or logicals. The most natural selection is that of positive integer vectors. The integers are indices of the elements to be selected and must be between 1 and **length**(x). The length of the index vector can vary:

```
> v <- 1:100
> v[6] # sixth element of v
> v[6:8] # 6th to 8th elements
> v[c(6,6,1:2)] # elements 6, 6, 1, and 2
> v[10:1] # elements 10, 9 ..., and 1
```

Another method is to remove the elements of the vector that you do not wish to retain: This is selection by negative integer vectors. The index vector indicates the indices of the elements to be excluded from the results:

```
> v[-(1:5)] # v without its five first elements
> v[-c(1,5)] # v without its first and its fifth elements
```

Another method is to select elements of the vector according to their value, or other elements originating from other R objects. This leads to selection by logical vectors. Using this selection method it is possible to extract specific elements which we already know how to characterise with a circumlocution, or by logical condition: "the element of sex M" or "the element with a value of less than 5 and/or more than or equal to 12":

```
> v <- 1:15
> print(v)
[1]  1  2  3  4  5  6  7  8  9 10 11 12 13 14 15
> v[(v<5)]
[1] 1 2 3 4
> v[(v<5)&(v>=12)] # & means "and"
integer(0)
```

As no elements are both less than 5 and greater than 12, the function yields an empty set with integer(0).

```
> v[(v<5)|(v>=12)]  # | means "or"
[1]  1  2  3  4 12 13 14 15
```

It is also possible to select the values of a vector from the values of another vector of the same length:

```
> T <- c(23, 28, 24, 32)
> O3 <- c(80, 102, 87, 124)
> O3[T>25]
[1] 102 124
```

1.4.4.5 Selection in Practice

We can substitute a selected subset with new values:

```
> x[is.na(x)] <- 0 # missing values (NA) are replaced by 0
> y[y<0]<- -y[y<0] #negative elts are replaced by their opposite
```

This action means taking the absolute value:

```
> y <- abs(y)
```

It is common to select the maximum or the minimum of a given function. Below are three possibilities for searching for the coordinate of the vector's smallest element x:

```
> which.min(x)
> which(x==min(x))
> (1:length(x))[x==min(x)]
```

The logical operator "==" is used to test equality and yields a Boolean.

1.4.5 Matrices

Matrices are atomic objects, that is to say, they are of the same mode or type for all values. Each value of the matrix can be located by its row and column numbers. The two intrinsic attributes of an R object are its **length**, which here corresponds to the total number of elements in the matrix, and its in the object **mode**, which here corresponds to the mode of the elements in this matrix. The matrices also include the dimension attribute **dim**, which yields the number of rows and the number of columns. They can also possess an optional attribute **dimnames** (see Section 1.4.7.4, p. 21). Hereafter are the main ways of creating a matrix. The most widely used is the **matrix** function which takes as its arguments the vector of elements and the number of rows or columns in the matrix:

```
> m <- matrix(c(1,17,12,3,6,0),ncol=2)
> m
     [,1] [,2]
[1,]    1    3
[2,]   17    6
[3,]   12    0
> m <- matrix(1:8,nrow=2)
> m
     [,1] [,2] [,3] [,4]
[1,]    1    3    5    7
[2,]    2    4    6    8
```

By default, R ranks the values in a matrix by column. To rank the elements by row, use the argument `byrow`:

```
> m <- matrix(1:8,nrow=2,byrow=TRUE)
> m
     [,1] [,2] [,3] [,4]
[1,]    1    2    3    4
[2,]    5    6    7    8
```

If the length of the vector is different from the number of elements in the matrix, R fills the whole matrix. If the vector is too large, it takes the first elements; and if the vector is too small, R repeats it:

```
> m <- matrix(1:4,nrow=3,ncol=3)
> m
     [,1] [,2] [,3]
[1,]    1    4    3
[2,]    2    1    4
[3,]    3    2    1
```

It is possible to fill a matrix with one single element without having to create the element vector:

```
> one <- matrix(1,nrow=2,ncol=4)
> one
     [,1] [,2] [,3] [,4]
[1,]    1    1    1    1
[2,]    1    1    1    1
```

Vectors are not considered by R to be matrices. However, it is possible to change a vector into a one-column matrix using the function **as.matrix**:

```
> x <- seq(1,10,by=2)
> x
[1] 1 3 5 7 9
> as.matrix(x)
     [,1]
[1,]    1
[2,]    3
[3,]    5
[4,]    7
[5,]    9
```

1.4.5.1   Selecting Elements or Part of a Matrix

The position of an element within a matrix is generally indicated by its row and column numbers. Thus, to select the element (i, j) from the matrix m, we write

```
> m[i,j]
```

It is unusual to only need to select one element in a matrix. Generally, multiple rows and/or columns are selected. Let us examine the different cases we might encounter in practice:

- Selection by positive integers:

```
> m[i,]
```

N.B. This yields row *i* in the form of a vector.

```
> m[i,,drop=F]
```

gives row *i* in the form of a one-row matrix and not of a vector, and thus the row name can be retained.

```
> m[,c(2,2,1)]
```

gives the second, the second and the first columns: This is a three-column matrix.

- Selection by negative integers:

```
> m[-1,]      # matrix m without its first row
> m[1:2,-1]   # two first rows of m without its first column
```

- Selection by logicals:

```
> m <- matrix(1:8,ncol=4,byrow=T)
> m
     [,1] [,2] [,3] [,4]
[1,]    1    2    3    4
[2,]    5    6    7    8
```

The following instruction yields only the columns of m for which the value on the first line is strictly greater than 2:

```
> m[,m[1,]>2]
     [,1] [,2]
[1,]    3    4
[2,]    7    8
```

This is of course a matrix, whereas the following instruction yields a vector containing the values of m which are greater than 2:

```
> m[m>2]
[1] 5 6 3 7 4 8
```

The following instruction replaces the values of m which are greater than 2 with NA:

```
> m[m>2] <- NA
> m
     [,1] [,2] [,3] [,4]
[1,]    1    2   NA   NA
[2,]   NA   NA   NA   NA
```

1.4.5.2   Calculating with the Matrices

```
> m <- matrix(1:4,ncol=2)
> m
     [,1] [,2]
[1,]    1    3
[2,]    2    4
> n <- matrix(3:6,ncol=2,byrow=T)
> n
     [,1] [,2]
[1,]    3    4
[2,]    5    6
> m+n
     [,1] [,2]
[1,]    4    7
[2,]    7   10
> m*n    # product element by element
     [,1] [,2]
[1,]    3   12
[2,]   10   24
> sin(m)  # sinus element by element
> exp(m) # exponential element by element
> m^4    # power 4  element by element
```

The following table gives the most useful functions for linear algebra.

Function	Description
X%*%Y	Product of matrices
t(X)	Transposition of a matrix
diag(5)	Identity matrix of order 5
diag(vec)	Diagonal matrix with the values of vector vec in the diagonal
crossprod(X,Y)	Cross product (t(X)%*%Y)
det(X)	Determinant of matrix X
svd(X)	Singular value decomposition
eigen(X)	Matrix diagonalisation
solve(X)	Matrix inversion

Function	Description
solve(A,b)	Solving linear systems
chol(Y)	Cholesky decomposition
qr(Y)	QR decomposition

Below is an example of how the functions **eigen** and **solve** are used:

```
> A <- matrix(1:4,ncol=2)
> B <- matrix(c(5,7,6,8),ncol=2)
> D <- A%*%t(B)
> D
     [,1] [,2]
[1,]   23   31
[2,]   34   46
> eig <- eigen(D)
> eig
$values
[1] 68.9419802  0.0580198

$vectors
           [,1]       [,2]
[1,] -0.5593385 -0.8038173
[2,] -0.8289393  0.5948762
```

The **eigen** function yields the eigenvalues in the object `values` and the eigenvectors in the object `vectors`. To extract the first eigenvector, simply write

```
> eig$vectors[,1]
```

For example, to solve the following system of equations:

$$\begin{cases} 23\ x + 31\ y = 1 \\ 34\ x + 46\ y = 2 \end{cases}$$

which is the same as writing `Dz=V`, with `V=c(1,2)`, we use the **solve** function

```
> V <- c(1,2)
> solve(D,V)
[1] -4  3
```

The solution is therefore $x = -4$ and $y = 3$.

1.4.5.3   Row and Column Operations

In this subsection we present a number of useful functions:

- Dimensions: **dim**(X), **nrow**(X) and **ncol**(X), respectively, give the dimension, the number of rows and the number of columns of X:

```
> X <- matrix(1:6,ncol=3)
> X
     [,1] [,2] [,3]
[1,]   1    3    5
[2,]   2    4    6
> ncol(X)
[1] 3
> nrow(X)
[1] 2
> dim(X)
[1] 2 3
```

These functions yield NULL if X is a vector.

- Concatenation: by column using the function **cbind**, by row using the function **rbind** (for further information, see Section 2.5, p. 43):

```
> cbind(c(1,2),c(3,4))
     [,1] [,2]
[1,]   1    3
[2,]   2    4
```

- The **apply** function is used to apply a function **f** to the rows (MARGIN=1) or the columns (MARGIN=2) of the matrix (for further information, see Section 4.2, p. 90). For example,

```
> apply(X,MARGIN=2,sum)  # sum by column
[1]   3  7 11
> apply(X,1,mean) # mean by row
[1] 3 4
```

- **sweep(X,2,stdev,FUN="/")** is used to divide each column (as MARGIN=2) of X by an element of the vector **stdev** (for further information, see Section 4.2, p. 93). This example is used to standardise the variables in a matrix.

Remark

Data cubes or higher dimensions tables can be constructed using the function **array**. These tables are manipulated in a similar way as the matrices.

1.4.6   Factors

Factors are vectors which enable the user to manipulate qualitative data. Length is determined by the **length** function, mode by **mode** and the categories of the factor by **levels**. They form a class of objects and are treated differently depending on the function, such as the **plot** function for graphs. Factors can be non-ordinal (male, female) or ordinal (level of skiing ability). Three functions can be used to create factors:

- The **factor** function

```
> sex <- factor(c("M","M","F","M","F","M","M","M"))
> sex
[1] M M F M F M M M
Levels: F M
```

We can also name each level as the factor is constructed:

```
> sex <- factor(c(2,2,1,2,1,2,1),labels=c("woman","man"))
> sex
[1] man man woman man woman man man man
Levels: woman man
```

- The **ordered** function

```
> ability <- ordered(c("beginner","beginner","champion",
 "champion","intermediate","intermediate","intermediate",
 "champion"),levels=c("beginner","intermediate","champion"))
> ability
[1] beginner      beginner     champion     champion
[5] intermediate intermediate intermediate champion
Levels: beginner < intermediate < champion
```

- The **as.factor** function

```
> somersault <- c(1:5,5:1)
> somersault
 [1] 1 2 3 4 5 5 4 3 2 1
> somersault.f <- as.factor(somersault)
> somersault.f
 [1] 1 2 3 4 5 5 4 3 2 1
Levels: 1 2 3 4 5
```

To know the levels, the number of levels and the number of factors per level of `somersault.f`, we use

```
> levels(somersault.f)
[1] "1" "2" "3" "4" "5"
> nlevels(somersault.f)
[1] 5
> table(somersault.f)
somersault.f
1 2 3 4 5
2 2 2 2 2
```

The **table** function can also be used to construct cross tabulations (Section 2.6, p. 46).

To convert the factors into numerics, we use the following instructions:

```
> x <- factor(c(10,11,13))
> as.numeric(x)
[1] 1 2 3
> as.numeric(as.character(x))
[1] 10 11 13
```

Details of this conversion are given in Section 2.3.1 (p. 35).

1.4.7   Lists

A list is a heterogeneous object. It is a set of ranked objects which do not always have the same mode or length. The objects are referred to as components of the list. These components can be named. Lists have the same two attributes as vectors (**length** and **mode**), as well as the additional attribute **names**. Lists are important objects as all the functions which yield multiple objects do so in the form of a list (see Section 1.5).

1.4.7.1   Creating a List

```
> myvector <- seq(2,10,by=3)
> mymatrix <- matrix(1:8,ncol=2)
> myfactor <- factor(c("M","M","F","M","F","M","M","M"))
> myordered <- ordered(c("beginner","beginner","champion",
  "champion","intermediate","intermediate","intermediate",
  "champion"),levels=c("beginner","intermediate","champion"))
> mylist <- list(myvector,mymatrix,myfactor,myordered)
> length(mylist)
[1] 4
> mode(mylist)
[1] "list"
```

We name the components of the list:

```
> names(mylist)
NULL
> names(mylist) <- c("vec","mat","sex","skilevel")
> names(mylist)
[1] "vec"   "mat"   "sex" "skilevel"
```

1.4.7.2   Extraction

To extract a component from the list, we can simply indicate the position of the element that we want to extract. `[[ ]]` are used to identify the element in the list:

```
> mylist[[3]]
[1] M M F M F M M M
Levels: F M
> mylist[[1]]
[1] 2 5 8
```

We can also use the name of the element, if it has one, which can be written in two ways:

```
> mylist$sex
[1] M M F M F M M M
Levels: F M

> mylist[["sex"]]
[1] M M F M F M M M
Levels: F M
```

It is possible to extract multiple elements from the same list, thus creating a sub-list. N.B. Here we use [ ] and not [[ ]]:

```
> mylist[c(1,3)]
$vec
[1] 2 5 8

$sex
[1] M M F M F M M M
Levels: F M
```

1.4.7.3 A Few List Functions

- **lapply** successively applies a function (such as the mean, the variance, etc.) to each of the components of a list.

- **unlist**(mylist) creates a single vector containing all the elements of the list. As all the elements of a vector are always of the same mode, automatic conversions in R must be used with caution.

- **c**(list1,list2) concatenates two lists.

1.4.7.4 Dimnames List

The dimnames list is an optional attribute of a matrix which contains two components: row names and column names. For example,

```
> X <- matrix(1:12,nrow=4,ncol=3)
> rowname <- c("row1","row2","row3","row4")
> colname <- c("col1","col2","col3")
> dimnames(X) <- list(rowname,colname)
> X
```

```
      col1 col2 col3
row1    1    5    9
row2    2    6   10
row3    3    7   11
row4    4    8   12
```

It is possible to select the rows and/or the columns using dimnames:

```
> X[c("row4","row1"),c("col3","col2")]
      col3 col2
row4   12    8
row1    9    5
```

It is also possible to modify (or delete) the row and/or column names using

```
> dimnames(X) <- list(NULL,dimnames(X)[[2]])
> X
      col1 col2 col3
[1,]    1    5    9
[2,]    2    6   10
[3,]    3    7   11
[4,]    4    8   12
```

1.4.8 Data-Frames

Data-frames are special lists with components of the same length but with potentially different modes. The data tables generally used in statistics are often referred to as data-frames. Indeed, a data table is made up of quantitative and/or qualitative variables taken from the same individuals.

The main ways of creating a data-frame are to use the **data.frame** function to concatenate vectors of the same size but which may have different modes; or the **read.table** function (see Section 2.1) to read a data table; the **as.data.frame** function for explicit conversion. For example, if we concatenate the numeric vector **vec1** and the alphabetic vector **vec2**:

```
> vec1 <- 1:5
> vec2 <- c("a","b","c","c","b")
> df <- data.frame(name.var1 = vec1, name.var2 = vec2)
> df
  name.var1 name.var2
1         1         a
2         2         b
3         3         c
4         4         c
5         5         b
```

In order to extract components from the data-frame, we can use the same methods as those presented for lists or matrices.

Remarks

- To convert a matrix to a data-frame, use the **as.data.frame** function.

- To convert a data-frame to a matrix, use the **data.matrix** function.

1.5 Functions

Functions are R objects. A great number of functions are predefined in R, for example, **mean** which calculates the mean, **min** which calculates the minimum, etc. It is also possible to create your own functions (see Section 4.3). Functions admit arguments as input and yield results as output.

1.5.1 Arguments of a Function

The arguments of a function can be compulsory or optional. When they are optional, they nonetheless have a default value.

Let us take the example of the function **rnorm** which generates random numbers according to the normal distribution. This function admits three arguments: n the number of values, mean the mean, and sd the standard deviation of the distribution. These arguments are by default set at 0 and 1. In order to be able to observe the same results from one simulation to another, you must "set the seed" of the random number generator using the function set.seed to a given value, here 13.

```
> set.seed(13) # specify generator seeds
> rnorm(n=3)
[1]   0.5543269 -0.2802719  1.7751634
```

These three numbers were thus chosen according to the standard normal distribution $\mathcal{N}(0,1)$. If we want to select four numbers according to a normal distribution with mean 5 and standard deviation 0.5, we use

```
> set.seed(13) # specify generator seeds
> rnorm(n=4,mean=5,sd=0.5)
[1] 5.277163 4.859864 5.887582 5.093660
```

It is not crucial to specify the name of the arguments, as long as the order is respected. This order is defined when the function is created. The **args** function yields the arguments of a function:

```
> args(rnorm)
function (n, mean = 0, sd = 1)
NULL
```

It is also possible to consult the help section for the function.

Some functions admit optional unnamed arguments. These are represented by ..., as for the **plot** function, for example,

```
> args(plot)
function (x, y, ...)
NULL
```

This function draws y according to x, with ... used to control the appearance of the graph (see Section 3.1.1, p. 52).

1.5.2 Output

Most of the functions yield multiple results and the entirety of these results is contained within a list. To visualise all these outputs, you first need to know all the elements. This information can be accessed using the **names** function. Use [[j]] to extract element *j* from the list, or indeed $ if the elements in the list are named (see Section 1.4.7, p. 20).

1.6 Packages

1.6.1 What Is a Package ?

A package or library of external programs is simply a set of R programs which supplements and enhances the functions of R. Packages are generally reserved for specific methods or fields of application.

There are more than 3000 packages (as of June 2011). A certain number of these are considered essential (MASS, rpart, etc.) and are provided with R. Others are more recent statistical advances and can be downloaded freely from the CRAN network. Furthermore, many more packages are available within projects, such as BioConductor which contains packages dedicated to processing genomic data.

1.6.2 Installing a Package

In this section we download a package and install it. Packages need only be installed once. Once they are installed, simply call the package using the **library** function each time you open an R session:

```
> library(package.name)
```

Packages are available on the CRAN network (http://cran.r-project.org/). A number of mirror sites, or exact copies of the CRAN site, are available. To install a package from CRAN, simply run the **install.packages** function:

> **install.packages**(dependencies=TRUE)

Then choose the nearest mirror to you and select the package to install, such as the FactoMineR package. You then simply need to load it in order to use it in an R session:

> **library**(FactoMineR)

It is also possible to download a package using one of the drop-down menus in Windows (Packages → Install package(s)...) or Mac OS X (Packages & Data → Package Installer). It is also possible to configure the use of a proxy (**help**(download.file)) and the destinations where the packages will be installed. All the packages installed on a given computer can be listed using the command

> **library**()

1.6.3   Updating Packages

Some packages constantly evolve, with new versions regularly available. It is therefore important to update them from time to time. To do this, they must not be in use. The best solution is to update packages when you open an R session by using

> **update.packages**()

to choose a mirror site near you and answer any questions.

In Windows and Mac OS X, there is a drop-down menu for updating packages: simply go to Packages → Update packages ... and choose a mirror site near you. A list of available package updates is then displayed, with all packages selected by default.

N.B. Package updates depend on the version of R used. Some packages are not compatible with past versions, or have not been compiled for these. We therefore also recommend that you regularly update your version of R before updating packages.

1.6.4   Using Packages

Once they are installed, packages must be called by the user before they can be used. Suppose we want to generate three vectors of $\mathbb{R}^2$ according to a multivariate normal distribution with mean $(0,1)$ and covariance matrix $\left(\begin{smallmatrix} 1 & 0.5 \\ 0.5 & 1 \end{smallmatrix}\right)$. This is possible thanks to the **mvrnorm** function from the MASS package. In order to use it, first call the MASS package and then use the function **mvrnorm**:

> **library**(MASS)
> **set.seed**(45) # specify generator seeds
> **mvrnorm**(3,mu=**c**(0,1),Sigma=**matrix**(**c**(1,0.5,0.5,1),2,2))

```
            [,1]        [,2]
[1,] -0.07788253 1.6681649
[2,] -1.05816423 0.8399431
[3,] -0.49608637 0.8387077
```

1.7   Exercises

Exercise 1.1 (Creating Vectors)
1. Create the following vectors using the **rep** function:

```
vec1 = 1 2 3 4 5 1 2 3 4 5 1 2 3 4 5
vec2 = 1 1 1 2 2 2 3 3 3 4 4 4 5 5 5
vec3 = 1 1 2 2 2 3 3 3 3 4 4 4 4 4
```

2. Create the following vector using the **paste** function:

```
vec4 = "A0)" "A1)" "A2)" "A3)" "A4)" "A5)" "A6)" "A7)"
       "A8)" "A9)" "A10)"
```

3. letters is a vector containing the 26 letters of the alphabet. Find the index of the letter q and create the vector containing "a1", "b2", $\cdots$ until "q" with the index of q.

Exercise 1.2 (Working with NA)
1. Fix the seed generator (**set.seed**) at 007 and create a vector vec1 of 100 numbers uniformly generated between 0 and 7 (**runif**). Calculate the mean and variance of vec1.

2. Create a vector vec2 equal to vec1 and select ten values at random (**sample**) and replace them by NA (missing values). Find the positions of the NA using the is.na function.

3. Calculate the mean and variance of the vector vec2.

4. Create a vector vec3 from vec2 by deleting the NA values and calculate the mean and variance.

5. Create a vector vec4 from vec2 where the NA values are replaced by the mean of the values of vec2. What happens to the mean and the variance of vec4?

6. Create a vector vec5 from vec2 where the NA values are replaced by values drawn from a Gaussian distribution with the mean and variance of vec2.

7. Same question as the previous one using values drawn from a Uniform distribution between the minimum and the maximum of the sample.

8. Create a vector vec7 from vec2 where the NA values are replaced by values selected at random from vec2.

Exercise 1.3 (Creating, Manipulating and Inverting a Matrix)

1. Create the following matrix mat (with the column and row names):

	column 1	column 2	column 3	column 4
row-1	1	5	5	0
row-2	0	5	6	1
row-3	3	0	3	3
row-4	4	4	4	2

2. Create a vector containing the diagonal elements of the matrix mat.

3. Create a matrix containing the first 2 rows of mat.

4. Create a matrix containing the last 2 columns of mat.

5. Calculate the determinant and then invert the matrix using the appropriate functions.

Exercise 1.4 (Selecting and Sorting in a Data-Frame)

1. From the iris dataset available in R (use **data(iris)** to load it and then iris[1:5,] to visualise the first five rows), create a sub-dataset comprising only the data for the category versicolor of the variable species (call this new dataset iris2).

2. Sort the data in iris2 in descending order according to the variable Sepal.Length (you can use the **order** function).

Exercise 1.5 (Using the apply Function)

1. Calculate the benchmark statistics (mean, min, max, etc.) for the three variables of the ethanol dataset (available in the lattice package).

2. Calculate the quartiles for each of the three variables. To do this, use the **apply** function with the **quantile** function.

3. Again using the **apply** function, calculate all the deciles for each of the three variables using the argument probs of the **quantile** function.

Exercise 1.6 (Selection in a Matrix with the apply Function)

1. From the matrix mat created in Exercise 1.3, create a matrix containing the columns of mat having all values smaller than 6 (use **apply** and **all**).

2. In the same way, create a matrix containing the rows of mat which do not contain 0.

Exercise 1.7 (Using the lapply Function)

1. Load the Aids2 dataset of package MASS and summarise the Aids2 dataset.

2. Using the **is.numeric** and the **lapply** functions, create a Boolean vector of length the number of columns of the data-frame `Aids2` which takes the value `TRUE` when the corresponding column of `Aids2` is not numeric and `FALSE` when it is numeric.

3. Select the qualitative variables of the `Aids2` data-frame and assign the result in a data-frame: `Aids2.qual`.

4. Give the levels of each qualitative variables of the `Aids2.qual` data-frame.

Exercise 1.8 (Levels of the Qualitative Variables in a Subset)
1. Load the `Aids2` dataset of package MASS.

2. Select the rows of `Aids2` where the sex is Male (`M`) and the state is not `Other`. Put the result in a new data-frame: `res`.

3. Summarise the `res` dataset and check that a) the `sex` variable has no female and b) the female level is still present.

4. Print the attributes of the `sex` variable (function **attributes**). Check that a) there is a level attribute with `F` and `M`, and b) the class of the object is factor.

5. Using **as.character**, transform the `sex` variable (of `res`) into an object named `sexc` of type character. Check that `sexc` has no attributes.

6. Using **as.factor**, transform `sexc` into an object named `sexf` of type factor. Check that the resulting object `sexf` has a) a level attribute, b) a class factor and c) no more `F` level.

Conclusion: Transforming a factor into a vector of characters and transforming back into a factor resets the levels.

7. Using Exercise 1.7, create a Boolean vector of length the number of columns of `res` which takes the value `TRUE` for categorical variables and `FALSE` for numeric variables.

8. Transform all factors of `res` into characters using **lapply**.

9. Transform back into factors using **lapply**.

2

Preparing Data

In statistics, data is the starting point for any analysis. It is therefore essential to know how to master operations such as importing and exporting data, changing types, identifying individuals with missing values, concatenating factor levels, etc. These different notions are therefore presented in this chapter, which aims to be both concise and satisfactory in practice.

2.1   Reading Data from File

The data is first collected, and may be pre-processed using software, be it a spreadsheet or statistical software. Each piece of software has its own storage format; the simplest option is to exchange data in a format common to all software, that is to say, in text format (extension .txt or .csv).

Data is generally contained within a file in which the individuals are presented in rows and the variables in columns. Text format contains data separated by a column separator. The table below thus groups together measurements of four variables (height, weight, shoe size, and sex) for three individuals, and the column separator used here is ",". The first column features the identifiers of the individuals, which here are their first names: all the first names must therefore be different. This identifier is generally a string of characters or the number of the individual. In some cases, individuals do not have identifiers and/or the variables do not have names.

```
surname,height,weight,size,sex
tony,184,80,9.5,M
james,175.5,78,8.5,M
john,158,72,8,M
```

Let us suppose that this data is contained within a mydata.csv file within a DATA file directory located in the current working directory (see Section 1.2.1 or 1.2.2 or 1.2.3). If inputting the data manually, make sure that you start a new line at the end of the last row; this operation is automatic in spreadsheets. The data is imported into the object tab as follows:

```
> tab <- read.table("DATA/mydata.csv",sep=",",header=TRUE,
          dec=".",row.names=1)
```

The file name is between speech marks (" or ' but not the same word-processing speech marks found in Microsoft® Word). The **sep** option indicates that the character which separates the columns is here " , ". For a space, use " ", and for a tabulation, "\t". The argument **header** indicates whether or not the first row contains the names of the variables. The argument **dec** specifies that the decimal point is a " . " (here for the number 175.5). Finally, the argument **row.names** indicates that column 1 is not a variable but the row identifier: here the individuals' names.

N.B. Windows users are accustomed to specifying paths using "\". With R, even in Windows, we use "/". Therefore, if the dataset is situated in the drive C: in the **Temp** directory, the reading becomes

```
> tab <- read.table("C:/Temp/mydata.csv",sep=",",header=TRUE,
        dec=".",row.names=1)
```

The path can also be a URL, as in the worked examples in the second part of this book. Thus, to import the dataset, it is possible to use the following command:

```
> decath <- read.table("http://www.agrocampus-ouest.fr/math/
    RforStat/decathlon.csv",sep=",",dec=".",header=TRUE,
    row.names=1)
```

Sometimes, a special character can be used to indicate that a value is missing. In R, this character is "NA" (Not Available). It is important to specify this character for missing data. Let us consider the following file mydata2.csv:

```
height weight size sex
184 80 9.5 "M"
175.5 78 8.5 "M"
. 72 8 "M"
178 . 7 "F"
```

It is imported by modifying the default value of **na.strings**, which is "NA", using

```
> tab <- read.table("DATA/mydata2.csv",sep=" ",header=TRUE,
        na.strings = ".")
```

The **quote** option (not used in our example) can be used to control any characters which may surround a string of characters. By default, this is a single or double speech mark. For other options used to control the reading more precisely, refer to the help of the **read.table** function.

It must be noted that the result of the reading is always a data-frame type object. The types of variables are inferred by the reading function, which can be wrong! It is therefore necessary to check each variable type using the **summary** function. In the **tab** table from the previous inputting, we obtain

```
> summary(tab)
      height          weight          size       sex
 Min.   :175.5   Min.   :72.00   Min.   :7.00   F:1
 1st Qu.:176.8   1st Qu.:75.00   1st Qu.:7.75   M:3
 Median :178.0   Median :78.00   Median :8.25
 Mean   :179.2   Mean   :76.67   Mean   :8.25
 3rd Qu.:181.0   3rd Qu.:79.00   3rd Qu.:8.75
 Max.   :184.0   Max.   :80.00   Max.   :9.50
 NA's   :  1.0   NA's   : 1.00
```

Here, the variable types are correct: height, weight and shoe size are quantitative and sex is qualitative with two categories (F and M). If this is not the case, it may be necessary to change the variable type (see Section 2.3.1). Sometimes, if the reading does not function correctly, the error may stem from the text file itself. Below is a list of some of the most common problems:

- Column separator incorrectly specified

- Decimal point incorrectly specified (the variables may be mistakenly considered qualitative)

- Tabulation replacing a space

- Row or column name includes an apostrophe

- Problem of open speech marks

In the first case, it is often difficult to find the mistake. One "unsophisticated" solution is offered in Exercise 2.1.

Remarks
- It is possible to import data using the Rcmdr package. We here advise you to turn to the appendix on Rcmdr (see Appendix A.3, p. 267).

- It is possible to import directly from databases such as access or mysql using the RMySQL package.

It must also be noted that certain (older) spreadsheets impose a limit of 256 columns. If the data has more than 256 variables, as in spectrometry or genomics, users of this type of spreadsheet should therefore present their datasets in inverted format: variables×individuals. In such cases, you must first transpose the table once you have imported it. This poses no further problems as long as all the variables are numerics. If even just one of the variables is qualitative, transposition will mean that there is a problem with variable type. Let us examine the following brief example:

```
sex M F M
age 18 17 22
```

This file was imported in the traditional manner, remembering that the first column contains the names of the variables:

```
> tab <- read.table("toyex.txt",sep=" ",header=F,row.names=1)
> summary(tab)
   V2        V3        V4
 18:1      17:1      22:1
 M :1      F :1      M :1
```

We transpose the table (the rows become the columns and vice versa):

```
> tab2 <- t(tab)
> summary(tab2)
 sex   age
 F:1   17:1
 M:2   18:1
       22:1
```

The **sex** variable poses no problem, but the **age** variable is summarised as a three-level factor. In fact, the result is no longer a data-frame but a character matrix.

```
> is.data.frame(tab2)
[1] FALSE
> is.matrix(tab2)
[1] TRUE
```

We must therefore convert our matrix to a data-frame (object which can contain both qualitative and quantitative variables) and then transform the second column into numerics. To do this, type the functions **as.character** and **as.numeric**:

```
> tab2 <- data.frame(tab2)
> tab2[,"age"] <- as.numeric(as.character(tab2[,"age"]))
> tab2
     sex   age
V2    M    18
V3    F    17
V4    M    22
```

A detailed explanation of the second row is given in Section 2.3.1, p. 35. Of course, if there are multiple numeric variables, they must all be checked, perhaps with the help of a loop (see Section 4.1.2, p. 87). To attribute a row heading other than V2, V3, V4, we use the **rownames** function. We can, for example, attribute the row names NULL (no names for the rows):

```
> rownames(tab2) <- NULL
> tab2
```

```
    sex  age
1    M   18
2    F   17
3    M   22
```

2.2 Exporting Results

Once the analyses have been conducted and the results obtained, it is often important to send the results to other people or export them to use with other software. For this purpose, a text file is again the most appropriate format. Results are mostly presented in table format and we will therefore export a table. Exporting a table is very simple:

```
> write.table(tab,"myfile.csv",sep=",",row.names=FALSE)
```

The object **tab** is exported to the file **myfile.csv** in the current working directory (see Section 1.2.1 or 1.2.2 or 1.2.3). We can of course specify a particular path. The exportation above can be used to control the column separator fixed at "," and the file name **myfile.csv**. It is also possible to control other options such as

- Whether or not the results contain the column names (**col.names**); by default **col.names= TRUE**

- Whether or not the results contain the row names (**row.names**); by default **row.names= TRUE**

- Whether or not character strings are defined by speech marks; by default **quote= TRUE**

- The decimal point (**dec**) which is "." by default and the character string for missing values (**na**) which is "NA" by default.

Therefore, to export without speech marks, without the row and column names, and with a tabulation separator, we use

```
> write.table(tab,"myfile.txt",quote=FALSE,row.names=FALSE,
    col.names=FALSE,sep="\t")
```

This type of file which mixes tabulation (separator) and spaces (missing value) is not recommended as it can be difficult to find the origin of the problem, should one occur. Judging from experience, the separator ";" seems the most appropriate choice.

Remarks
- The **write.csv** function is a possible alternative to the **write.table** function.

- The **write.infile** function from the FactoMineR package can be used to write all the objects of a list in the same file, without having to specify each of the objects in the list.

2.3 Manipulating Variables

2.3.1 Changing Type

It is often necessary to change the variable type. For example, following inputting, a qualitative variable whose categories are coded using figures is understood by R to be a quantitative variable. In such cases, they must be changed from quantitative (type: `numeric`) to qualitative (type: `factor`).

Converting quantitative variables to qualitative variables is easy: simply create a factor using the **factor** function. Let us take the example of a variable X made up of numerics:

```
> X <- c(rep(10,3),rep(12,2),rep(13,4))
> X
[1] 10 10 10 12 12 13 13 13 13
```

There are classical methods for knowing whether a vector-type object is a quantitative or a qualitative variable without displaying the entire vector. The first way of doing this is to ask R about the type:

```
> is.factor(X)
[1] FALSE
> is.numeric(X)
[1] TRUE
```

The second is to summarise the variable (**summary**). When it is a quantitative variable, the minimum, the maximum, the quartiles and the mean are displayed. However, for a factor, the number of observations for the first six levels of the qualitative variable is given:

```
> summary(X)
   Min. 1st Qu.  Median    Mean 3rd Qu.    Max.
  10.00   10.00   12.00   11.78   13.00   13.00
```

This, of course, is a quantitative variable. This is the most practical instruction for data tables (data-frames) which have been imported. Indeed, the summary is offered variable by variable and enables the user to quickly identify any errors of variable type.

Converting to a factor is just as simple; use the **factor** function:

```
> Xfac <- factor(X)
```

```
> Xfac
[1] 10 10 10 12 12 13 13 13 13
Levels: 10 12 13
> summary(Xfac)
10 12 13
 3  2  4
```

Displaying a factor enables the user to distinguish it from a numeric, thanks to the presence of different `levels` at the end of the display. This is also the case for the summary yielded by **summary**.

Converting a factor to a numeric is a two-step process. We first convert the factor to a character vector and then convert this vector to a numeric. If we convert the factor directly to a numeric, the levels will be re-coded in order (the first level will be 1, the second 2, etc.):

```
## coerce to factor + label change
> as.numeric(Xfac)
[1] 1 1 1 2 2 3 3 3 3
## coerce to factor without label change: 2 steps
> foo <- as.character(Xfac)
> foo
[1] "10" "10" "10" "12" "12" "13" "13" "13" "13"
> as.numeric(foo)
[1] 10 10 10 12 12 13 13 13 13
```

2.3.2   Dividing into Classes

Changing a quantitative variable to a qualitative variable, or dividing it into classes, is common practice in statistics. Thus, for example, to construct the axes of a Multiple Correspondence Analysis with quantitative variables (see Worked Example 10.3, p. 230), they must be converted to qualitative variables. Continuous variables such as X can be divided in one of two ways:

- Division according to thresholds defined by the user

- Automatic division into similar-sized classes

Let us start with the first option. We assign X a vector of fifteen random numbers following a standard normal distribution:

```
> set.seed(654) ## set generator seeds
> X <- rnorm(15)
> X
 [1] -0.76031762 -0.38970450  1.68962523 -0.09423560  0.095301
 [7]  1.06576755  0.93984563  0.74121222 -0.43531214 -0.107260
[13] -0.98260589 -0.82037099 -0.87143256
```

Let us divide this variable into three levels: between the minimum and −0.2, between −0.2 and 0.2, and finally between 0.2 and the maximum. The **cut** function can be used to conduct this step automatically. The classes are of the format $(a_i; a_{i+1}]$. In order to include the minimum of X in the first class, you must use the argument `include.lowest=TRUE`. The first class will therefore be of the form $[a_1, a_2]$:

```
> Xfac <- cut(X,breaks=c(min(X),-0.2,0.2,max(X)),
      include.lowest=TRUE)
> Xfac
 [1] [-0.983,-0.2] [-0.983,-0.2] (0.2,1.69]    (-0.2,0.2]
 [5] (-0.2,0.2]    (0.2,1.69]    (0.2,1.69]    (0.2,1.69]
 [9] (0.2,1.69]    [-0.983,-0.2] (-0.2,0.2]    [-0.983,-0.2]
[13] [-0.983,-0.2] [-0.983,-0.2] [-0.983,-0.2]
Levels: [-0.983,-0.2] (-0.2,0.2] (0.2,1.69]
> table(Xfac)
Xfac
[-0.983,-0.2]    (-0.2,0.2]    (0.2,1.69]
            7             3             5
```

The result of the division is a factor. When we list observations by category (with the **table** function), we obtain unequal class sizes. If we require similar class sizes for each of the three categories, we use the **quantile** function. This function is used to calculate empirical quantiles and thus gives the required thresholds. The thresholds are as follows:

- The minimum, which is the empirical quantile associated with the probability 0: $\mathbb{P}_n(X < \min(X)) = 0$

- The first threshold, which is the empirical quantile associated with the probability 1/3: $\mathbb{P}_n(X \leq \text{threshold}_1) = 1/3$

- The second threshold, which is the empirical quantile associated with the probability 2/3: $\mathbb{P}_n(X \leq \text{threshold}_2) = 2/3$

- The maximum, which is the empirical quantile associated with the probability 1: $\mathbb{P}_n(X \leq \max(X)) = 1$

```
> mybreaks <- quantile(X,probs=seq(0,1,length=4))
> Xfac <- cut(X,breaks=mybreaks,include.lowest=TRUE)
> table(Xfac)
Xfac
[-0.983,-0.544]    (-0.544,0.311]    (0.311,1.69]
              5                 5               5
```

2.3.3   Working on Factor Levels

This section helps to resolve some of the classical problems when it comes to organising data:

- How can multiple factor levels be merged if, for example, they have insufficient samples?

- How can the levels of a factor be renamed?

The answer to these two questions is the same in R. Let us use the variable `Xfac` from the previous section:

```
> Xfac
 [1] [-0.983,-0.544] (-0.544,0.311]  (0.311,1.69]
 [4] (-0.544,0.311]  (-0.544,0.311]  (0.311,1.69]
 [7] (0.311,1.69]    (0.311,1.69]    (0.311,1.69]
[10] (-0.544,0.311]  (-0.544,0.311]  [-0.983,-0.544]
[13] [-0.983,-0.544] [-0.983,-0.544] [-0.983,-0.544]
Levels: [-0.983,-0.544] (-0.544,0.311] (0.311,1.69]
```

We want to change the titles of the levels. To do this, simply assign a new value to the levels of `Xfac` called using the **levels** function:

```
> levels(Xfac) <- c("lev1","lev2","lev3")
> Xfac
 [1] lev1 lev2 lev3 lev2 lev2 lev3 lev3 lev3 lev3
[10] lev2 lev2 lev1 lev1 lev1 lev1
Levels: lev1 lev2 lev3
```

To merge two levels such as levels 1 and 3, simply rename the two levels with the same name (i.e. repeat it):

```
> levels(Xfac) <- c("lev1+3","lev2","lev1+3")
> Xfac
 [1] lev1+3 lev2   lev1+3 lev2   lev2   lev1+3 lev1+3 lev1+3
 [9] lev1+3 lev2   lev2   lev1+3 lev1+3 lev1+3 lev1+3
Levels: lev1+3 lev2
```

The order in which the levels of a factor appear is not insignificant. For ordinal factors, simply specify the order of the levels when the factor is created (Section 1.4.6, p. 18). In theory this order is not important for factors whose categories are not ordinal. However, in certain methods such as analyses of variance (see Worked Example 8.1, p. 157), the first level is used as a base level, and the others are all compared to it. In such cases, specifying the first factor is thus essential. Les us consider, for example, a qualitative variable with levels 1, 2 or 3, depending on whether the individual received the `classic`, `new` or `placebo` treatment. The data is as follows:

```
> X <- c(1,1,2,2,2,3)
```

and to convert it to a factor:

```
> Xfac <- factor(X,label=c("classic","new","placebo"))
> Xfac
[1] classic classic new     new     new     placebo
Levels: classic new placebo
```

If we want `placebo` to be the base level, then we use the **relevel** function:

```
> Xfac2 <- relevel(Xfac,ref="placebo")
> Xfac2
[1] classic classic new     new     new     placebo
Levels: placebo classic new
```

In order to control the order of the levels, simply recreate a factor by modifying an existing one, specifying the order in which the levels should be displayed. Thus, to choose the order of levels `placebo` then `new` and finally `classic`, we use

```
> Xfac3 <- factor(Xfac,levels=c("placebo","new","classic"))
> Xfac3
[1] classic classic new     new     new     placebo
Levels: placebo new classic
```

Sometimes we have to delete atypical individuals[1] or work with a subset of individuals. This may mean that there could be qualitative variables for which a level has no individuals. Let us construct an example to illustrate this problem:

```
> myfactor <- factor(c(rep("A",3),"B",rep("C",4)))
> myfactor
[1] A A A B C C C C
Levels: A B C
```

If we need to remove the fourth individual, the factor is as follows:

```
> myfactor <- myfactor[-4]
> myfactor
[1] A A A C C C C
Levels: A B C
> table(myfactor)
myfactor
A B C
3 0 4
```

The category `B` thus no longer has any individuals. We must therefore delete this category. In order to do so, it is possible to reassign this category to another, such as category `A`. Another method is to convert the factor into a

---

[1] "The last mass trials were a great success. There are going to be fewer but better Russians." Greta Garbo in *Ninotchka* (1939), by Ernst Lubitsch.

character vector and then to convert it back to a factor. The first conversion removes the notion of levels, and the second returns it, taking into account the data vector. As category B no longer exists, it disappears:

```
> myfactor <- as.character(myfactor)
> myfactor <- factor(myfactor)
> myfactor
[1] A A A C C C C
Levels: A C
```

Sometimes, when a category's sample size is too small, we break down the individuals into other categories at random. This process, known as ventilation (random allocation), is examined in Exercise 2.6.

2.4 Manipulating Individuals

2.4.1 Identifying Missing Data

In R, missing data is represented by NA (Not Available). In order to identify the individuals for which they are missing, simply use the **is.na** function, which yields TRUE if the value is NA and FALSE if it is not. Let us construct a variable measured from ten individuals by conducting a random draw according to normal distribution $\mathcal{N}(0, 1)$. Then, we assign missing data, for individuals 3, 4 and 6 for example:

```
> set.seed(23)
> variable <- rnorm(10,mean=0,sd=1)
> variable[c(3,4,6)] <- NA
```

Now let us imagine that the statistician needs to know how to identify those individuals with missing values. His first task will be to discover the identifiers, or numbers for these individuals. In order to find them, simply use the **is.na** function:

```
> miss <- is.na(variable)
> miss
 [1] FALSE FALSE  TRUE  TRUE FALSE  TRUE FALSE FALSE FALSE FALSE
```

Elements 3, 4 and 6 yield the value TRUE. They therefore carry the value NA in the `variable` vector. The **which** function automatically yields the indices of TRUE:

```
> which(miss)
[1] 3 4 6
```

If we want to eliminate these individuals, we must conserve all those which are not missing, that is to say, all the FALSE of the miss vector. To do this, we conserve all the TRUE for negations of miss, that is to say !miss:

```
> variable2 <- variable[!miss]
> variable2
[1]  0.19321233 -0.43468211  0.99660511 -0.27808628  1.01920549
[6]  0.04543718  1.57577959
```

Another way to do this is to eliminate the coordinates using selection by negative integers:

```
> variable3 <- variable[-which(miss)]
> variable3
[1]  0.19321233 -0.43468211  0.99660511 -0.27808628  1.01920549
[6]  0.04543718  1.57577959
```

Let us extend this to a data table which admits multiple variables. We create a second variable and construct the table don:

```
> myfac <- factor(c(rep("M",3),NA,NA,rep("F",5)))
> tab <- cbind.data.frame(variable,myfac)
> tab
      variable myfac
1    0.19321233     M
2   -0.43468211     M
3           NA     M
4           NA  <NA>
5    0.99660511  <NA>
6           NA     F
7   -0.27808628     F
8    1.01920549     F
9    0.04543718     F
10   1.57577959     F
```

The statistical summary of the dataset is

```
> summary(tab)
    variable          myfac
 Min.   :-0.4347   F   :5
 1st Qu.:-0.1163   M   :3
 Median : 0.1932   NA's:2
 Mean   : 0.4454
 3rd Qu.: 1.0079
 Max.   : 1.5758
 NA's   : 3.0000
```

The total of missing individuals is specified variable by variable. If we want to eliminate those individuals with at least one missing value, we must first identify them:

```
> miss <- is.na(tab)
> miss
      variable varqual
 [1,]    FALSE   FALSE
 [2,]    FALSE   FALSE
 [3,]     TRUE   FALSE
 [4,]     TRUE    TRUE
 [5,]    FALSE    TRUE
 [6,]     TRUE   FALSE
 [7,]    FALSE   FALSE
 [8,]    FALSE   FALSE
 [9,]    FALSE   FALSE
[10,]    FALSE   FALSE
```

Next, the rows of `tab` which correspond to the rows of `miss` with at least one TRUE must be eliminated: here the rows 3, 4, 5 and 6. In order to do this, for each row we need to know whether there is at least one TRUE. The **any** function applied to a vector yields TRUE if there is at least one TRUE in the vector. To apply this function to every row in the table, we use **apply** on the rows (MARGIN=1):

```
> deletion <- apply(miss,MARGIN=1,FUN=any)
> deletion
 [1] FALSE FALSE TRUE TRUE TRUE TRUE FALSE FALSE FALSE FALSE
```

The last vector therefore contains TRUE in each row which contains at least one NA. The rows to be conserved are therefore those which correspond to all of the TRUE of the negation of `deletion`. The refined data is therefore

```
> tab2 <- tab[!deletion,]
> tab2
      variable myfac
1   0.19321233     M
2  -0.43468211     M
7  -0.27808628     F
8   1.01920549     F
9   0.04543718     F
10  1.57577959     F
```

Finally, if we want to know the row × column pairs of individuals with a missing value, we can use the option **arr.ind** (for array indices) of the **which** function:

```
> which(is.na(tab),arr.ind=TRUE)
      row col
 [1,]   3   1
 [2,]   4   1
```

```
[3,]   6   1
[4,]   4   2
[5,]   5   2
```

The first piece of missing data is in row 3 and column 1 ..., the fifth piece of missing data is in row 5 and column 2.

2.4.2   Finding Outliers

We use data offered by R to illustrate this problem. The dataset is called kyphosis and is part of the rpart package (see also Section 1.6). Load the package and then the dataset to the session in order to be able to use them:

```
> library(rpart)
> data(kyphosis)
```

We use the Number variable from this dataset. Let us draw a boxplot (Figure 2.1) for this variable:

```
> boxplot(kyphosis[,"Number"])
```

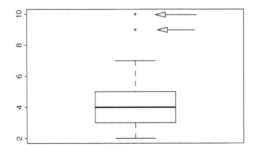

Figure 2.1
Boxplot for a variable with outliers (denoted by arrows).

The two individuals above the higher whisker (see Figure 2.1) are often considered to be outliers. How can we determine their number (or other identifier)? One solution is to use the result of the **boxplot** function. The component out from the ouput list gives us the outliers' values:

```
> results <- boxplot(kyphosis[,"Number"])
> outliers <- results$out
> outliers
[1]   9 10
```

The outlier values for the variable Number are therefore 9 and 10. We still need to know the identifier for the individuals in question. In order to do this, we search for individuals whose value corresponds to one of the values contained within `outliers`. The operator `%in%` yields `TRUE` or `FALSE`, depending on whether or not the value is within the set (here `outliers`). Then, to know the number of observations which are found within this set, we use **which**:

```
> which(kyphosis[,"Number"]%in%outliers)
[1] 43 53
```

Individuals 43 and 53 are therefore the outliers.

Another way of identifying these outliers is to use the **identify** function. By clicking on a point on a graph, the function yields the row number corresponding to this individual. The **identify** function takes the abscissa and ordinate values from the scatterplot as the arguments:

```
> identify(rep(1,length(kyphosis[,"Number"])),kyphosis[,"Number"])
```

Here we can see that only one variable is represented on the ordinate. The abscissa is simply 1 as we only have one variable for all the observations. The number of observations is here **length(kyphosis[,"Number"])**.

We can therefore identify the outliers with a single click. When you have finished, simply click elsewhere in the graph window with the right button of the mouse, or type Esc in the command window.

2.5 Concatenating Data Tables

Let us imagine that we have two data tables that we want to merge. This can be done in one of two ways: putting the tables one on top of the other (juxtaposition of rows, see Figure 2.2) or putting them one next to the other (juxtaposition of columns, see Figure 2.3).

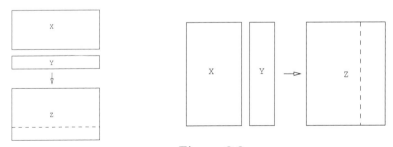

Figure 2.3
Column concatenation: **cbind(X,Y)**.

Figure 2.2
Row concatenation: **rbind(X,Y)**.

In order to concatenate by row for the tables X and Y, we use

```
> Z <- rbind(X,Y)
```

X and Y must therefore have the same number of columns. However, the logic is a little different, depending on whether we concatenate two matrices or two data-frames. This question is illustrated in the example below.

Let us first construct a matrix with its row identifiers and variable names X1 and X2:

```
> X <- matrix(c(1,2,3,4),2,2)
> rownames(X) <- paste("row",1:2,sep="")
> colnames(X) <- paste("X",1:2,sep="")
> X
     X1 X2
row1  1  3
row2  2  4
```

And the same for the Y matrix:

```
> Y <- matrix(11:16,3,2)
> colnames(Y) <- paste("Y",1:2,sep="")
> Y
     Y1 Y2
[1,] 11 14
[2,] 12 15
[3,] 13 16
```

We concatenate X and Y in one single matrix:

```
> Z <- rbind(X,Y)
> Z
     X1 X2
row1  1  3
row2  2  4
     11 14
     12 15
     13 16
```

In this example, R concatenates the matrices conserving only the variable names of the first matrix.

If we use two data-frames with identical structures, an error message appears when the variable names are different:

```
> Z <- rbind(data.frame(X),data.frame(Y))
Error in match.names(clabs, names(xi)) :
  names do not match previous names
```

We must therefore change the variable names for one of the data-frames:

```
> Xd <- data.frame(X)
> Yd <- data.frame(Y)
```

```
> colnames(Yd) <- c("X2","X1")
> rbind(Xd,Yd)
      X1 X2
row1  1   3
row2  2   4
3     14 11
4     15 12
5     16 13
```

We can see (on the third line) that the order of the variables is inconsistent. In this case, the variables are re-classified prior to concatenation.

Concatenation by column is done in the same way as for the matrices or data-frames. The **cbind** function does not check the row names. The names from the first table are retained:

```
> X <- matrix(c(1,2,3,4),2,2)
> rownames(X) <- paste("row",1:2,sep="")
> Y <- matrix(11:16,2,3)
> cbind(data.frame(X),Y)
     X1 X2 X1 X2 X3
row1  1  3 11 13 15
row2  2  4 12 14 16
```

This type of concatenation – that is to say, by row or by column – is not the only possibility. Two tables can be merged according to a key, **merge** (see Figure 2.4).

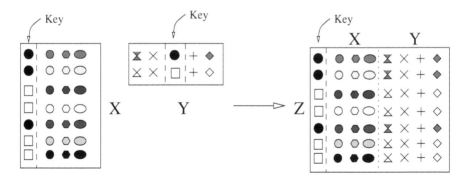

Figure 2.4
Merging by key: **merge(X,Y,by="key")**.

Let us construct two data tables, one with more rows than the other. The first table merges (column concatenation) a continuous variable (**age**) and two qualitative variables in a data-frame. The **cbind.data.frame** function can be used to specify that the result must be a data-frame:

```
> age <- c(7,38,32)
> surname <- c("arnaud","nicolas","laurent")
> town <- factor(c("rennes","rennes","marseille"))
> indiv <- cbind.data.frame(age,surname,town)
> indiv
  age surname      town
1   7  arnaud    rennes
2  38 nicolas    rennes
3  32 laurent marseille
```

A second table groups together the characteristics of the towns:

```
> population <- c(200,500,800)
> towns <- cbind.data.frame(c("rennes","lyon","marseille"),
      population)
> names(towns) <- c("town","pop")
> towns
        town pop
1     rennes 200
2       lyon 500
3  marseille 800
```

If we want to merge these two tables into one, where the characteristics of the towns will be repeated in each row, we merge by the key `town`:

```
> merge(indiv,towns,by="town")
       town age surname pop
1 marseille  32 laurent 800
2    rennes   7  arnaud 200
3    rennes  38 nicolas 200
```

The key `town` is therefore the first column of the dataset.

2.6   Cross-Tabulation

When two qualitative variables are analysed, we can present the data in two ways: a classical table where an individual (a row) is described by the variables (two variables), or a cross-tabulation (also known as a contingency table), which gives the sample size for each confrontation of categories. Of course, this can be generalised to more than two qualitative variables.

In order to get a clear idea of both types of table, let us first construct two qualitative variables (`wool` and `tension`) collected from ten individuals. The variable `wool` corresponds to the following three types of wool: Angora, Merinos and Texel. The variable `Tension` indicates the value `Low` or `High` for the wool's tension resistance.

```
> tension <- factor(c(rep("Low",5),rep("High",5)))
> tension
 [1] Low  Low  Low  Low  Low  High High High High High
Levels: High Low
> wool <- factor(c(rep("Mer",3),rep("Ang",3),rep("Tex",4)))
> wool
 [1] Mer Mer Mer Ang Ang Ang Tex Tex Tex Tex
Levels: Ang Mer Tex
```

Let us merge these two variables in the table don:

```
> tab <- cbind.data.frame(tension,wool)
> tab
   tension wool
1      Low  Mer
2      Low  Mer
3      Low  Mer
4      Low  Ang
5      Low  Ang
6     High  Ang
7     High  Tex
8     High  Tex
9     High  Tex
10    High  Tex
```

The cross-tabulation is therefore obtained directly on the objects using

```
> table(tab$tension,tab$wool)
       Ang Mer Tex
  High   1   0   4
  Low    2   3   0
```

Another method is to use the column names from the data-frame and to conduct the confrontation using a formula:

```
> crosstab <- xtabs(~tension+wool,data=tab)
> crosstab
        wool
tension Ang Mer Tex
   High   1   0   4
   Low    2   3   0
```

Many functions in R are based on the hypothesis of an individuals × variables table, but the opposite operation must also be made possible for situations in which the data are contained within a cross-tabulation. However, an individuals × variables table is not constructed immediately:

```
> tabframe <- as.data.frame(crosstab)
```

```
> tabframe
  tension wool Freq
1    High  Ang    1
2     Low  Ang    2
3    High  Mer    0
4     Low  Mer    3
5    High  Tex    4
6     Low  Tex    0
```

We obtain the frequencies for each combination rather than a row for each individual. To reconstruct the initial table, see Exercise 2.8.

2.7   Exercises

Exercise 2.1 (Robust Reading of Data)
In some cases, reading data fails and it can be difficult to understand why. In such cases, here is a more robust procedure which can be used to identify the problem.

1. Read the file mydata.csv in a character vector myvector using the **scan** function (in a complex reading, this step can also be used to check the column separator).

2. Change the decimal separator from "," to "." (function **gsub**) in myvector. It must be noted that this step can be conducted more simply using a spreadsheet or word processor.

3. Construct a character matrix mymatrix of four rows and five columns which contains the data and the names of the individuals and variables.

4. Gather the names of individuals and variables in an object namerow and namecol. Allocate the remaining data within a three-row and four-column matrix (mymatrix). For a complex or delicate inputting, display the data column by column in order to make sure that the columns are not "mixed up".

5. Convert the matrix mymatrix to a data-frame mydata and check that the variable type is correct. If not, convert the factors into numerics. See Section 2.3.1, p. 35.

Exercise 2.2 (Reading Data from File)
Read the files test1.csv, test1.prn, test2.csv and test3.csv.

Exercise 2.3 (Reading Data from File with Date Format)
1. Read the file test4.csv (Hint: Use the argument **skip** to delete the first 2

lines). The data presents a sample of friends with the age, gender (coded 0 for male and 1 for female) and the first date of skiing (year-month-day).

2. Read the file again with the option `colClasses` to specify that the gender should be considered as a factor and the first time skiing as a date.

Exercise 2.4 (Reading Data from File and Merging)
1. Read the files `state1.csv`, `state2.csv` and `state3.csv`.

2. Merge the three data-frames into one using the common columns to repeat the rows adequately.

Exercise 2.5 (Merging and Selection)
1. Read the files `fusion1.xls` and `fusion2.xls`.

2. Retain only the variables `yhat1`, `yhat3`, `Rhamnos`, and `Arabinos`. With these four variables, create one single data-frame.

3. Create the variable `yres1`, which is the difference between `yhat1` and `Rhamnos`, and create the variable `yres2`, which is the difference between `yhat3` and `Arabinos`. Add these variables to the data-frame.

Exercise 2.6 (Ventilation (allocation at random))
In the presence of the following qualitative variable,

```
> Xfac <- factor(c(rep("A",60),rep("B",20),rep("C",17),
    rep("D",3)))
```

1. Calculate the frequencies of each category (or factor level).

2. Display the names of the categories with sample sizes smaller than 5% of the total sample size.

3. Calculate the frequencies of each category without the category or categories of the previous question. The result will be placed in the `freq` vector.

4. Select the individuals which take the category or categories of the Exercise 2.6.2. Allocate them a value selected from the remaining categories, according to a draw for which the probabilities are calculated in Exercise 2.6.3 (use the **sample** function). This process is known as ventilation[2].

Exercise 2.7 (Ventilation by Ordinal Factors)
In the presence of the following qualitative variable with ordinal categories:

---

[2] "Have you ever seen anything like that? Totally calm, singing away and then bang! Right in the face! This man is crazy! And I treat madmen every day; I'll write him a prescription, and a strong one! I'll show him who's boss. We'll find bits of him all around Paris like pieces of a jigsaw puzzle... When it goes too far, I don't correct people any more, I destroy them... Spread them out, ventilate them...". Bernard Blier in *Crooks in Clover*, by Georges Lautner (1963), script by Michel Audiard.

```
> Xfac <- factor(c(rep("0-10",1),rep("11-20",3),rep("21-30",5),
     rep("31-40",20),rep("41-50",2),rep("51-60",2),
     rep("61-70",1),rep("71-80",31),rep("+ 80",20)))
```

1. Calculate the frequencies of each category (or factor level).

2. Display the names of the categories with sample sizes smaller than 5% of the total sample size.

3. Start with the "weakest" (according to the order of the categories) of the categories in Exercise 2.7.2. Merge this category with the category immediately "above". If the sample size of these two merged categories is greater than 5% of the total, go on to the next category of Exercise 2.7.2 (if there is one). If not, merge again with the category immediately "above" (out of all the categories). Proceed in the same way until all the categories from Exercise 2.7.2 have been used.

Exercise 2.8 (Cross-Tabulation → Data Table)
1. Create the contingency table confronting the qualitative variables types of wool (wool) and tension (tension):

```
       Ang Mer Tex
High    1   0   4
Low     2   3   0
```

2. From this table, create a character matrix tabmat with three columns and as many rows as confrontations of categories (or cells in the contingency table). This character matrix will be filled row by row with tension (rows from the previous table), wool type (columns from the previous table), and the sample size for the confrontation of categories. In order to do so, we use the functions **matrix** and **rep**.

3. Convert the character matrix from the previous question to a data-frame; name it tabframe, for example, and check the variable types. Using this data-frame, and assign the total number of individuals to n (**sum**) and the number of qualitative variables to nbfac (**ncol**).

4. Create a counter iter set at 1 and a character matrix tabcomplete, for example the character "", the same size as the final dataset.

5. Make a loop around the number of rows of tabmat (see Section 4.1.2, p. 87). Each row i corresponds to a confrontation of categories (or strata). If the number of individuals with this confrontation of categories i is not null, then repeat in as many rows as necessary in tabcomplete, the confrontation of category i (by distributing the categories in the corresponding columns). It is possible to use a loop, the iter counter, the tabmat matrix and the tabframe data-frame.

The result tabcomplete will be the same as the initial table.

3

R Graphics

Graphs are often the starting point for statistical analysis. One of the main advantages of R is how easy it is for the user to create many different kinds of graphs. We begin this chapter by studying conventional graphs, followed by an examination of some more complex representations. This final part, using the lattice package, can be omitted if using R for the first time.

3.1 Conventional Graphical Functions

To begin with, it may be interesting to examine a few example of graphical representations which can be constructed with R. We use the **demo** function:

```
> demo(graphics)
```

It must be noted that there are other demonstrations available, a list of which can be found using **demo()**.

We give examples of representations for quantitative and qualitative variables. We use the data file `ozone.txt`, imported using

```
> ozone <- read.table("ozone.txt",header=T)
```

We obtain meteorological variables and an ozone pollution variable collected in Rennes (France) during the summer of 2001. The variables available are `maxO3` (maximum daily ozone), `T12` (temperature at midday), `wind` (direction), `rain` and `Wx12` (projection of the wind speed vector on the east-west axis at midday):

```
> ozone <- ozone[,c("T12","maxO3","wind","rain","Wx12")]
> summary(ozone)
      T12              maxO3            wind         rain
 Min.   :14.00   Min.   : 42.00   East :10   Dry  :69
 1st Qu.:18.60   1st Qu.: 70.75   North:31   Rainy:43
 Median :20.55   Median : 81.50   South:21
 Mean   :21.53   Mean   : 90.30   West :50
 3rd Qu.:23.55   3rd Qu.:106.00
 Max.   :33.50   Max.   :166.00
```

```
        Wx12
Min.    :-7.878
1st Qu.:-3.565
Median :-1.879
Mean    :-1.611
3rd Qu.: 0.000
Max.    : 6.578
```

The variables T12, max03 and Wx12 are continuous quantitative variables (numerics), whereas wind and rain are factors.

3.1.1   The Plot Function

The **plot** function is a generic function (see Appendix A.1, p. 257) of R used to represent all kinds of data. Classical use of the **plot** function consists of representing a scatterplot for a variable y according to another variable x. For example, to represent the graph of the function $x \mapsto \sin(2\pi x)$ on $[0, 1]$, at regular steps we use the following commands:

```
> gridx <- seq(0,1,length=50)
> fx <- sin(2*pi*gridx)
> plot(x=gridx,y=fx)
```

These commands open a graph window (known as a device).

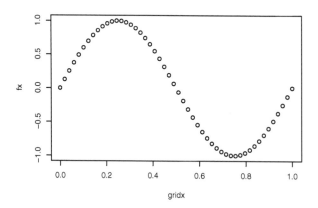

Figure 3.1
Scatterplot for the function $x \mapsto \sin(2\pi x)$.

In statistics it is more common to use the **plot** function with formulae of the type y~x. This is the format that we will use. For example, the graph in Figure 3.1 can also be obtained using

```
> plot(fx~gridx)
```

We continue to use the ozone example and represent different pairs of variables. Let us start by representing two quantitative variables: maximum ozone (max03) according to temperature (T12) (Figure 3.2):

```
> plot(ozone[,"max03"]~ozone[,"T12"])
```

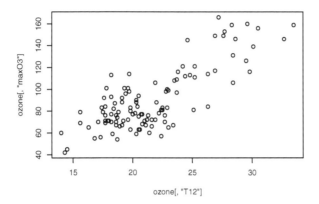

Figure 3.2
Scatterplot (T12, max03).

As the two variables are contained and named within the same table, a simpler syntax can be used, which automatically inserts the variables as labels for the axes (Figure 3.3):

```
> plot(max03~T12,data=ozone)
```

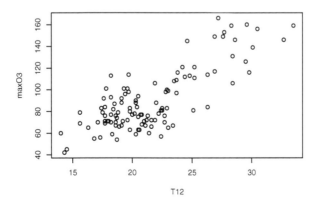

Figure 3.3
Scatterplot (T12, max03) with explicit axis labels.

This graphical representation could also have been obtained using the command:

```
> plot(ozone[,"max03"]~ozone[,"T12"],xlab="T12",ylab="max03")
```

Similarly, to represent a quantitative variable (`max03`) according to a qualitative variable (`wind`), we write

```
> plot(max03~wind,data=ozone,xlab="Wind directions",
    ylab="Maximum Ozone Concentration")
```

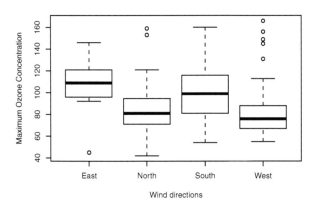

Figure 3.4
Boxplots of `max03` according to `wind`.

In this case, the **plot** function yields a boxplot for each category of the variable `wind` (Figure 3.4). In this graph we can see if the variable `wind` has an effect on the ozone. Here, the easterly wind seems to be associated with high concentrations of ozone. This graph can also be obtained using the **boxplot** function:

```
> boxplot(max03~wind,data=ozone)
```

Given the nature of the variables, the **plot** function chose to use the **boxplot** function as it considers it the most relevant.

We can also represent two qualitative variables using a bar chart. For example,

```
> plot(rain~wind,data=ozone)
```

can be used to obtain, for each category of the explanatory factor (here `wind`), the relative frequencies for each category of the response factor (here `rain`). The width of the bar is proportional to the frequency of the category of the explanatory factor (here `wind`). This graph can also be obtained using the **spineplot** function.

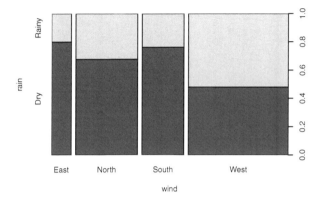

Figure 3.5
Bar chart for `rain` according to `wind`.

In Figure 3.5, we find that the weather type "rain" is proportionally greater when the wind blows from the west (which indeed seems to be the case in Rennes). Furthermore, as this is the widest bar, we can also conclude that this is the most common wind direction in Rennes.

Finally, it is possible to represent a qualitative variable according to a quantitative variable, here `wind` according to T12:

```
> plot(wind~T12,data=ozone)
```

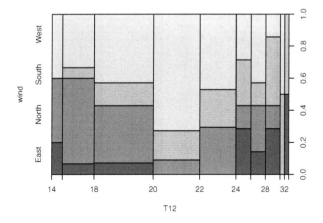

Figure 3.6
Bar chart for `wind` according to T12.

We here obtain a bar chart (Figure 3.6). On the abscissa, the quantitative variable is divided into classes in the same default manner as a histogram (**hist**). In each of these classes, the relative frequencies of each category of `wind` are calculated, giving the height of each bar. The colour corresponds to the category. This graph can also be obtained automatically using

```
> spineplot(wind~T12,data=ozone)
```

This bar chart is difficult to interpret. It is possible to return to the first formulation of the **plot** function to represent the scatterplot (T12,wind); see Figure 3.7:

```
> plot(ozone[,"T12"],ozone[,"wind"],yaxt="n",
    xlab="T12",ylab="wind",pch="+")
> axis(side=2,at=1:4,labels=levels(ozone[,"wind"]))
```

It must here be noted that the graph must be reworked: The argument `yaxt="n"` deletes the scale from the y axis, which here takes the different categories of the qualitative variable converted to numerics by default: 1 to 4. The **axis** function is therefore used to rename the scale.

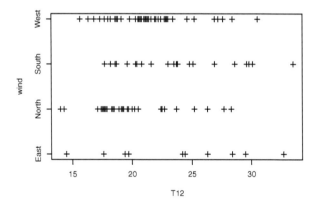

Figure 3.7
Graph of the qualitative variable `wind` according to the quantitative variable T12.

This type of graph is interesting when there are only a few measurements for each category. If not, the data should be represented using a boxplot (see Figure 3.4).

Even if, intrinsically, the **plot** function expects the two input arguments `x` and `y` for the abscissa and ordinate, respectively, it is possible to specify only one argument. For example, for one quantitative variable, R draws this variable sequentially with the observation number on the abscissa axis (see Figure 3.8).

```
> plot(ozone[,"maxO3"],xlab="num.",ylab="maxO3",cex=.5,pch=16)
```

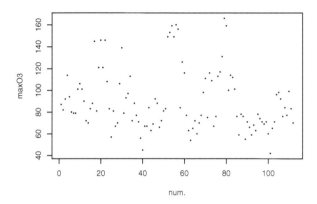

Figure 3.8
Scatterplot for ozone peaks (maxO3) according to index.

This command yields an image where, by default, the abscissa axis carries the ordinal continuation $1, \cdots, n$ known as "index" (where n is the number of observations). The size of the symbol can be controlled using the argument cex which manages the increase (or decrease) factor for size (by default cex=1). The argument pch is used to specify the shape of the points. This argument accepts numerical values between 0 and 25 (see Figure 3.9) or a character directly rewritten on the screen.

Figure 3.9
Symbol obtained for the value of the argument pch.

It is also possible to modify the type of line drawn using the argument type: "p" to only draw the points (default option), "l" (line) to link them, "b" or "o" to do both. The argument type can also be used to draw vertical bars ("h", high-density) or steps after the points ("s", step) or before ("S", step). A graphical illustration of these arguments is given on Figure 3.10.

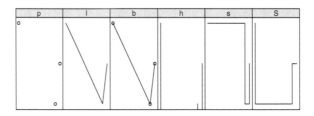

Figure 3.10
Argument `type` for the values "p", "l", "b", "h", "s", "S".

The evolution of ozone maxima can be obtained using the argument `type="l"` (Figure 3.11):

```
> plot(ozone[,"maxO3"],xlab="num.",ylab="maxO3",type="l")
```

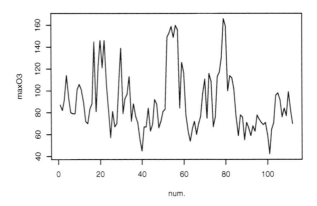

Figure 3.11
Evolution of ozone peaks (`maxO3`) during the summer of 2001.

This is a very classical graph for representing temporal data.

In conclusion, the **plot** function adapts its behaviour to each type of data.

3.1.2   Representing a Distribution

To represent the distribution of a continuous variable (e.g. the vector of numerics `ozone[,"maxO3"]`), there are a number of pre-programmed classical solutions. The histogram for the variable `maxO3` (Figure 3.12) estimates the density if we specify the argument `prob=TRUE` of the **hist** function:

```
> hist(ozone[,"maxO3"],main="Histogram",prob=TRUE,xlab="Ozone")
```

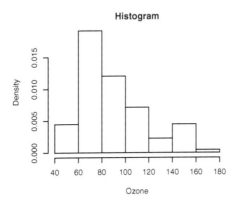

Figure 3.12
Histogram for ozone.

A kernel density estimate of the variable `maxO3` can be obtained using the **density** function. The graphical representation (Figure 3.13) is obtained as follows:

```
> plot(density(ozone[,"maxO3"]),main="Kernel Density Estimate",
    xlab="Ozone")
```

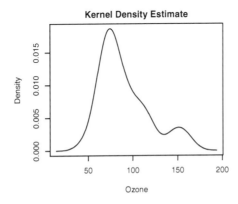

Figure 3.13
Kernel density estimate for ozone.

To represent the distribution of a qualitative variable, we use a bar chart, obtained using the **barplot** function. We must count the number of occurrences at each level using the **table** function. For the factor `wind`,

```
> barplot(table(ozone[,"wind"]))
```

R yields a bar chart (Figure 3.14) with, on the x-axis the values of the factor, and on the y-axis the number of observations for each category.

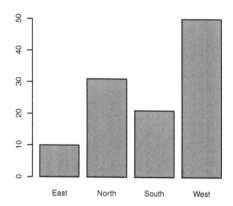

Figure 3.14
Bar chart for a qualitative variable.

It is also possible to use the **plot** function directly on the vector of the categories:

```
> plot(ozone[,"wind"])
```

Another method, but which is not recommended, is circular representation, or a pie chart, which can be obtained using the **pie** function.

3.1.3   Adding to Graphs

Once the graph is drawn, it is possible to add additional information: extra lines (**lines**), points (**points**), texts (**text**), symbols (circles, squares, stars, etc.; **symbols**), arrows (**arrows**), segments (**segments**) or polygons (**polygon**). We will only present the addition of texts, lines and points.

We again refer to the graphical representation of ozone maxima `max03` according to temperature `T12` (Figure 3.15). To make the date appear we need to add text (**text**) to the scatterplot obtained using **plot**.

Nonetheless, in order to avoid overlapping the symbols (the small circle `pch=1`) and the names, we do not draw any symbols (`type="n"`). We simply write the month and the day (i.e. the last four characters of the row names) with a reduced font size (`cex=.75`):

```
> plot(maxO3~T12,data=ozone,type="n")
> text(ozone[,"T12"],ozone[,"maxO3"],substr(rownames(ozone),5,8),
    cex=.75)
```

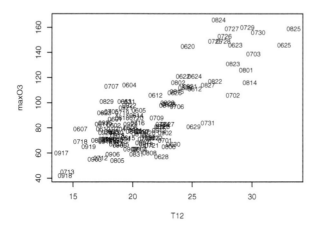

Figure 3.15
Text positioned via the points' coordinates.

The subsequent graph (Figure 3.15) does not show the exact position of the name. It is therefore recommended to plot crosses on the graph and then to add text above these crosses. We thus draw these crosses on the coordinates (**plot** function), and then add the text on top of that (pos=3) offset by 0.3 (offset):

```
> plot(maxO3~T12,data=ozone,type="p",pch=3,cex=.75)
> text(ozone[,"T12"],ozone[,"maxO3"],
    substr(rownames(ozone),5,8),cex=.75,pos=3,offset=.3)
```

If in addition we wish to add a coloured symbol for the dates corresponding to rainy days, we will select their coordinates and then draw a grey dot at their coordinates (Figure 3.16):

```
> selection <- ozone[,"rain"]=="Rainy"
> points(ozone[selection,"T12"],ozone[selection,"maxO3"],
    pch=21,bg="grey70",cex=.75)
```

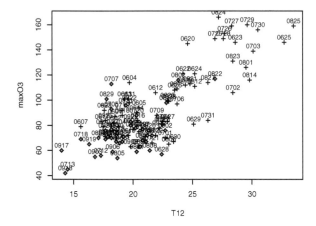

Figure 3.16
Scatterplot with the rainy days in grey.

From this graph we can see that it does not rain when the temperature exceeds 27 degrees. This limit can be marked using the **abline** function and adding the command (Figure 3.17):

> **abline(v=27,lty=2)**

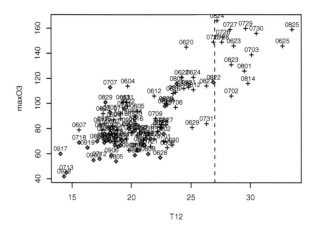

Figure 3.17
Use of the **abline** function.

We can also add a horizontal line using **abline(h=)** or any type of line (by specifying the intercept and the slope using **abline(c(intercept,slope))**.

The lty (line type) setting is used to control the type of line (1: continuous line, 2: dashed, 3: dotted, etc.).

As its name indicates, the **abline** function adds a straight line onto the existing graph. If we want to add two broken lines to compare, for example, the evolution of maximum ozone levels over two different weeks, we use the **lines** function. Let us compare the evolution of ozone levels during the first two weeks:

```
> plot(ozone[1:7,"maxO3"],type="l")
> lines(ozone[8:14,"maxO3"],col="grey50") # add lines (2nd week)
```

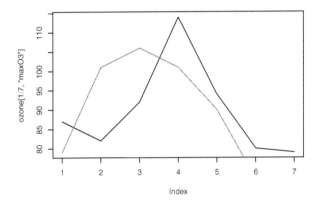

Figure 3.18
Adding a grey line using **lines**.

The graph (Figure 3.18) does not show the sixth and seventh observations for this second week. Indeed, if we want to draw a line on the graph, we need to know the minimum and maximum for each axis in order to be able to prepare them and to prepare the scales. Here, the minimum for the second line is less than that of the first. The graph is scaled using the **plot** function. When called upon, this function is only aware of the information concerning the **plot** instruction. Scaling is not automatic following a **lines** (or **points**) order. It is therefore important, right from the **plot** order, to specify the minimum and maximum for both weeks. Simply give the ylim argument the results of the **range** function. The latter yields the minimum and maximum. This argument tells the **plot** function between which ordinates the graph is situated (Figure 3.19).

```
> rangey <- range(ozone[1:7,"maxO3"],ozone[8:14,"maxO3"])
> plot(ozone[1:7,"maxO3"],type="l",ylim=rangey,lty=1)
> lines(ozone[8:14,"maxO3"],col="grey50",lty=1)
```

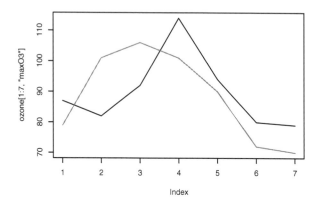

Figure 3.19
Adding a grey line using **lines** and scaling the ordinates using the `ylim` argument.

Of course, if the abscissas are the same, you do not need to specify the minimum and maximum on the abscissa.

3.1.4   Graphs with Multiple Dimensions

The most classical functions for representing graphs in 3D on a grid of points are the functions **persp** (3D with perspective effect), **contour** (contour lines) and **image** (contour lines in colour).

The classic "Mexican hat", given in the help section of **persp**, is the graph of the function $(x, y) \mapsto z = f(x, y) = 10 \ \sin\left(\sqrt{x^2 + y^2}\right)/\sqrt{x^2 + y^2}$. To implement this function, we have chosen a regular square grid with thirty different values between -10 and 10 (on x and y):

```
> f <- function(x,y)  10 * sin(sqrt(x^2+y^2))/sqrt(x^2+y^2)
> x <- seq(-10, 10, length= 30)
> y <- x
```

We evaluate the **f** function at each point of the grid: for each pair (`x[i]`, `y[j]`), we calculate `f(x[i],y[j])`. In order to avoid creating a (double) loop, we use the **outer** function which enables this type of evaluation:

```
> z <- outer(x, y, f)
```

We draw this function in 3D (Figure 3.20)

```
> persp(x, y, z, theta = 30, phi = 30, expand = 0.5)
```

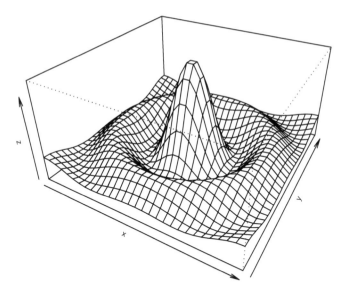

Figure 3.20
Example of use of the **persp** function.

We could obtain level sets using **contour**(x, y, z) or **image**(x, y, z).

We can also use the rgl package to construct the response surface for the previous graph:

```
> library(rgl)
> rgl.surface(x, y, z)
```

This package is recommended for all problems with 3D visualisation. For example, it can be used to rotate graphs with the mouse and to use symbols with lighting effects.

If we want to draw a scatterplot representing n individuals characterised by three quantitative variables, it is not possible to use "conventional" graphs. We must therefore use a package such as rgl or lattice. We start with the lattice package which is one of the default packages provided with R.

We again use the ozone example, representing the individuals for the three quantitative variables max03, T12 and Wx12. As we are looking to explain the variable max03, it is only natural to represent it on the third axis (Figure 3.21):

```
> library(lattice)
> cloud(max03~T12+Wx12,type=c("p","h"),data=ozone)
```

Here, the lattice package enables the user to apply two arguments for type, which is not possible in "conventional" graphs.

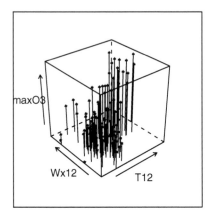

Figure 3.21
3D graph characterised by measurements of the quantitative variables max03,
T12 and Wx12.

If we want to use the rgl package, simply type:

```
> plot3d(ozone[,"T12"],ozone[,"Wx12"],ozone[,"maxO3"],radius=2,
     xlab="T12",ylab="Wx12",zlab="maxO3",type="s")
```

It must be noted that the type of points type="s", specific to rgl, enables
spheres to be drawn with a fixed radius of 2, thanks to the argument radius.

3.1.5 Exporting Graphs

Graphs are used as a visual support and are often needed for use in presentations or other types of external documents. It is therefore important to be able
to export them in a format which is compatible with other software. There
are many different formats available, such as jpeg, png, pdf and postscript
(ps), or vectorial graphical formats such as emf (in Windows), svg or xfig.

A device is defined as either the graph window or a file into which the
graph is exported. The device changes with the export format (see Table
3.1).

In Windows, it is easiest to save the graph from the graph window. In order
to do so, go to File → Save as and choose the desired format, preferably a
vector format (metafile "emf"). It is also possible to copy the graph and to
paste it into a document by right-clicking on the graph with the mouse and
then choosing Copy as metafile.

Otherwise, exporting a graph always follows the same procedure. For
example, if we want to export in pdf, the name of the device is (pdf) and is
activated by the function of the same name. To save the graph Figure 3.19 in
pdf format, write

TABLE 3.1

Non-exhaustive List of Devices

Device	Description	Graph Window
quartz	default Mac OS X device	Yes
X11	Unix/Linux default device	Yes
jpeg	jpeg file	No
png	png file	No
bmp	bmp file	No
pdf	pdf file	No
ps	ps file	No
pictex	picTeX macros	No
svg	svg vector format	No
xfig	xfig vector format	No

```
> pdf("graphik.pdf")
> rangey <- range(ozone[1:7,"maxO3"],ozone[8:14,"maxO3"])
> plot(ozone[1:7,"maxO3"],type="l",ylim=rangey,lty=1)
> lines(ozone[8:14,"maxO3"],col="grey50",lty=1)
> dev.off()
```

The **dev.off** function closes the device and finalises the file `graphik.pdf` which will be found in the current working directory (see Section 1.2.1 or 1.2.2 or 1.2.3). It is of course possible to specify the destination pathway using `pdf("mypath/graphik.pdf")`.

There are a number of different options when exporting graphs (paper size, choice of fonts, etc.). These options all depend on the export format chosen and can be consulted in the help section. Not all the devices are necessarily available: they depend on the platform used, the compilation options and the packages installed.

If we want to modify a graph (move a legend, text, increase the size of a given element, etc.), vector graphics software is required. The graph must therefore be exported (format svg, emf or xfig, for example) and then altered using the appropriate software. It must be noted that inkscape and oodraw (from openoffice) are freely available open source software which import graphs in svg format. In Mac OS X and Unix/Linux, xfig is another alternative to these two pieces of software.

3.1.6 Multiple Graphs

If we want to feature more than one graph in the same window, after opening the device, simply state the number of graphs required. Traditionally, we use the **par** function. The instruction **par(mfrow=c(n,p))** organises np graphs into n rows and p columns. To obtain two graphs on the same row (Figure 3.22), write:

```
> par(mfrow=c(1,2))
> plot(1:10,10:1,pch=0)
> plot(rep(1,4),type="l")
```

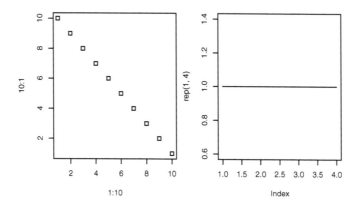

Figure 3.22
Graphs arranged in 1 row × 2 columns using **par(mfrow=c(1,2))**.

To come back to one graph per window, simply kill the graph window or use the command **par(mfrow=c(1,1))**.

Sometimes, we need to use graphs of different sizes. In such cases, we use the **layout** function: this function divides the graph window into areas before distributing them. As its argument, it therefore admits a nrow × ncol matrix, which creates nrow × ncol number of individual tiles. The values of the matrix therefore correspond to the numbers of the graphs featured in each tile. For example, to set out three graphs on two rows, as in Figure 3.23, we first create the following matrix:

$$\begin{pmatrix} 1 & 1 \\ 2 & 3 \end{pmatrix}$$

```
> mat <- matrix(c(1,1,2,3), nrow=2, ncol=2, byrow = TRUE)
```

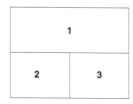

Figure 3.23
Positioning three graphs on two rows using **layout**.

Then we draw the three graphs (Figure 3.24):

```
> layout(mat)
> plot(1:10,10:1,pch=0)
> plot(rep(1,4),type="l")
> plot(c(2,3,-1,0),type="b")
```

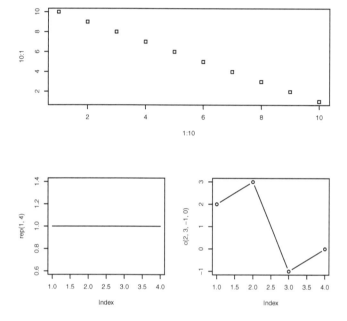

Figure 3.24
Three graphs on two rows using **layout**.

It is also possible to specify the height and width of the columns (see help for **layout** and Exercise 3.8). It must be noted that the graphs might be (too) far apart, but this can easily be resolved (see Exercise 3.8).

3.1.7   Multiple Windows

Sometimes, more than one graph window is required. To create a graph window, simply open the device which corresponds to the operating system: **X11** in Unix/Linux or Windows and **quartz** in Mac OS X. Therefore, if we want to draw up two graphs in two separate windows, in Unix/Linux or Windows simply use

```
> plot(1:10,10:1,pch=0)
> X11()
> plot(rep(1,4),type="l")
```

The first graph is thus drawn. If no graph window was open prior to the creation of the first graph, it will be created automatically. A second window is then opened, thus rendering the first inactive. The second graph is drawn in this new window. Subsequent graphs will be drawn in the active window.

3.1.8   Improving and Personalising Graphs

In order to improve a graph, it is possible

- To use colours:

```
> plot(x,y,type="p",pch=o,col="red")
```

These colours can be specified using a number (generally, 1=black, 2=red, 3=green, etc.), a name in English from the list given by the vector **colors**(), or an RGB code with transparency (see **rgb**). The correspondence between the number of the colour and the colour itself is ensured by the palette. It is possible to change the background and line colours for a device using the **par** function, which manages all the graphical parameters for a given device:

```
> par(fg="blue",bg="#f2a1c2")
```

- To modify the palette, ensuring a correspondence between the number and the colour:

```
> palette(gray(seq(0,0.9,len=10))) # palette of grey
> plot(1:10,rep(1,10),type="p",pch=25,bg=1:10)
> palette("default")
```

- To add a title:

```
> plot(x,y,type="p",pch=o,main="My Title")
```

- To control the appearance of the axes or to delete them (and the corresponding legend):

```
> plot(c(10,11),c(12,11),type="p",axes=FALSE,xlab="",ylab="")
```

to finally "redefine them by hand":

```
> axis(1,at=c(10,11),label=c("Dim 1","Dim 2"))
```

- To construct orthonormal axes (argument asp=1):

```
> plot(x,y,type="p",asp=1)
```

- To add a legend:

```
> rangey <- range(ozone[1:7,"maxO3"],ozone[8:14,"maxO3"])
> plot(ozone[1:7,"maxO3"],type="l")
> lines(ozone[8:14,"maxO3"],ylim=rangey,col="grey50"))
> legend("topleft",legend=c("week 1","week 2"),
      col=c("black","grey50"),lty=1)
```

- To insert symbols or mathematical formulae (**help**(plotmath)), for example for the legend of an axis:

```
> plot(1,1,xlab=expression(bar(x)==sum(frac(x[i], n), i==1,n)))
```

Some parameters can be modified directly in the graph command, and others are accessible using the **par** function.

Argument	Description
adj	Controls the position of the text in relation to the left edge of the text: 0 to the left, 0.5 centred, 1 to the right; if two values are given **c(x,y)**, the text will be horizontally and vertically justified
asp	Specifies the ratio between y and x: asp=1 constructs orthonormal graphs
axes	By default TRUE, the axes and the frame are shown
bg	Specifies the background colour 1, 2, or a colour chosen from **colors**()
bty	Controls the frame line, permitted values: "o","l", "7", "c", "u" or "]" (with the shape of frame resembling that of the character); bty="n" deletes the frame
cex	Controls the size of the characters and symbols compared to the default value which is 1
cex.axis	Controls the size of the characters on the scales of the axes
cex.lab	Controls the size of the characters for the axis labels
cex.main	Controls the size of the characters of the title
cex.sub	Controls the size of the characters in the caption
col	Specifies the colour of the graph: possible values 1, 2, or a colour chosen from **colors**()
col.axis	Specifies the colour of the axes
col.main	Specifies the colour of the title
font	Controls the style of the text (1: normal, 2: italics, 3: bold, 4: bold italics)
font.axis	Controls the style for the scales
font.lab	Controls the style for the axis labels
font.main	Controls the style of the title
font.sub	Controls the style of the caption

Argument	Description
`las`	Controls the distribution of annotations on the axes (0, default value: parallel to the axes, 1: horizontal, 2: perpendicular to the axes, 3: vertical)
`lty`	Controls the type of line drawn (1: continuous, 2: dashed, 3: dotted, 4: alternately dotted and dashed, 5: long dashes, 6: alternately long and short dashes), or write in words "solid", "dashed", "dotted", "dotdash", "longdash", "twodash" or "blank" (to write nothing)
`lwd`	Controls the thickness of the lines
`main`	Specifies the title of the graph, for example `main="My Title"`
`mfrow`	Vector `c(nr,nc)` which divides the graph window into `nr` rows and `nc` columns; the graphs are then drawn in a line
`offset`	Specifies the position of the text with respect to the point position (**text** function)
`pch`	Integer (between 0 and 25) which controls the type of symbol (or potentially any character in speech marks)
`pos`	Specifies the position of the text; permitted values 1, 2, 3 and 4 (**text** function)
`ps`	Integer controlling the size of the text and symbols in points
`sub`	Specifies the caption of the graph, for example `sub="My Caption"`
`tck, tcl`	Specifies the distances between graduations on the axes
`type`	Specifies the type of graph drawn: permitted values `"n","p","l","b,"h","s","S"`
`xlim, ylim`	Specifies the limits of the axes, for example `xlim=c(0,10)`
`xlab, ylab`	Specifies the annotations on the axes, for example `xlab="Abscissa"`

Below is a list of classical graphical functions.

Function	Description
barplot(x)	Draws a bar chart for the values of x
boxplot(x)	Draws a boxplot for x
contour(x,y,z)	Draws contour lines, see also `filled.contour(x,y,z)`
coplot(x~y|f1)	Draws the bivariate plot of (x, y) for each value of f1 (or for each small interval for the values of f1)
filled.contour(x, y,z)	Draws contour lines, but the areas between the contours are coloured, see also `image(x,y,z)`

Function	Description
hist(x,prob=TRUE)	Draws a histogram for x
image(x,y,z)	Draws rectangles at the coordinates x, y coloured according to z, see also contour(x,y,z)
interaction.plot(f1, f2,x,fun=mean)	Draws the graph for the averages of x according to the values of f1 (on the abscissa axis) and f2 (one broken line per category of f2)
matplot(x,y)	Draws the bivariate plot for the first column of x confronted with the first column of y, the second column of x with the second column of y, etc.
pairs(x)	If x is a matrix or data-frame, draws all the bivariate plots between the columns of x
persp(x,y,z)	Draws a response surface in 3D, see demo(persp)
pie(x)	Draws a pie chart
plot(objet)	Draws a graph corresponding to the class of objet
plot(x,y)	Draws y against x
qqnorm(x)	Draws the quantiles of x in relation to those expected from a normal distribution
qqplot(x,y)	Draws the quantiles of y in relation to those of x
spineplot(f1,f2)	Draws the bar chart corresponding to f1 and f2
stripplot(x)	Draws the graph for the values of x on a line
sunflowerplot(x,y)	Idem, but the overlapped points are drawn in the shape of flowers with the number of petals corresponding to the number of points
symbols(x,y,...)	For the coordinates x and y, draws different symbols (stars, circles, boxplots, etc.)

3.2 Graphical Functions with **lattice**

The lattice package is used to represent those data for which the variable represented on the ordinate axis is based on multiple variables. To do this, the lattice package relies on low-level graphical functions which are not those we traditionally use, but instead are those offered in the grid package.

The grid package is more comprehensive and more accurate than older low-level graphical functions used, for example, in **plot**. However, they are more complex and we refer the interested reader to the book by Murrel (2005). Here we present a brief overview of the possibilities offered by the lattice package.

We use the quakes dataset available in R. The dataset gives the locations of 1000 seismic events of MB greater than 4.0. The events occurred near Fiji since 1964. These include the Richter magnitude magn, longitude long, latitude lat, and the depth depth. The spatial distribution of the earthquakes provides information about the location of the trench. The most natural way of representing this is to feature the longitude on the abscissa, the latitude on the ordinate, and to plot one point for each earthquake. As it is probable that the location will affect the depth of the earthquake, we have chosen to represent a symbol with a colour which darkens as the depth increases. To do this, we divide the depth variable into six classes (see Section 2.3.2, p. 35):

```
> library(lattice)
> data(quakes)
> breakpts <- quantile(quakes$depth,probs=seq(0,1,length=7))
> class <- cut(quakes$depth,breaks=breakpts,include.lowest=TRUE)
> levels(class) <- c("grey80","grey65","grey50","grey35",
      "grey10","black")
> plot(lat ~ long,data=quakes,col=as.character(class),pch=16)
```

Quakes seem therefore to be distributed according to two planes (see Figure 3.25). We can represent them according to six depth levels using the

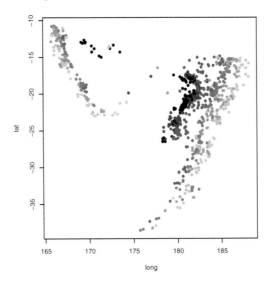

Figure 3.25
Latitude and longitude of earthquakes near Fiji.

functions **xyplot** and **equal.count**. The sample sizes for each level are equal and the intervals overlap: 5% of the data are common to two consecutive intervals (`overlap=0.05`).

```
> Depth <- equal.count(quakes$depth,number = 6,overlap=0.05)
> xyplot(lat ~ long | Depth, data = quakes,pch="+",
      xlab = "Longitude",ylab = "Latitude")
```

The depth interval is given in the strip above each panel of the overall graph (Figure 3.26).

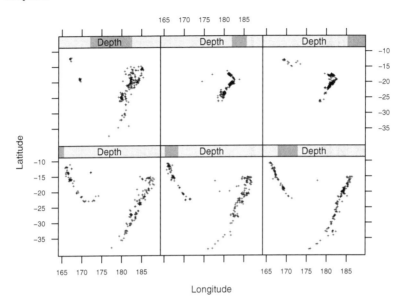

Figure 3.26
Latitude and longitude of earthquakes depending on depth.

3.2.1 Characteristics of a "Lattice" Graph

Below are some of the characteristics of graphs from the lattice package:

- In general, each graph is composed of a series of panels (elementary graphs) arranged into a rectangular table.

- Panels are indexed from left to right and from bottom to top.

- The abscissa and ordinate axes are identical from one panel to the next.

- A panel is a graph of one variable (ordinate) in relation to another (abscissa) conditioned by one (or multiple) other variable(s). The conditional value appears in the strip for each panel.

"Lattice" type graphs are similar to normal graphs but they can be used to divide groups of variables. Table 3.2 illustrates the correspondence between conventional and "lattice" graphs.

Conditioning relies on a qualitative variable or on the way a continuous variable is partitioned. In the latter case, the resulting object is referred to as "shingles". This object is created using the **equal.count** function and can be represented graphically (Figure 3.27) using the generic **plot** function:

```
> plot(Depth)
```

TABLE 3.2

Correspondence between Conventional and "Lattice" Graphs

Lattice	Description	Conventional
barchart	Bar chart	**barplot**
bwplot	Boxplot	**boxplot**
densityplot()	Estimation of density	**plot(density())**
histogram	Histogram	**hist**
qqmath	Normal QQ plot	**qqnorm**
qq	Quantiles - quantiles (2 samples)	**qqplot**
xyplot	Scatterplot	**plot**
levelplot	Greyscale (3D)	**image**
contourplot	Contour lines (3D)	**contour**
cloud	Scatterplot (3D)	none
wireframe	Perspective (3D)	**persp**
splom	Matrix of scatterplot	**pairs**
parallel	Diagonal	none

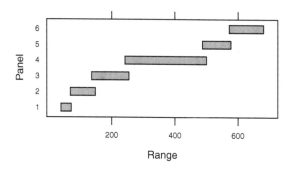

Figure 3.27
Shingles of earthquake depth.

3.2.2 Formulae and Groups

"Lattice" graphs are based on a formula which can be used to specify the abscissa and ordinate axes and the conditioning. For example, if within the same data-frame, we have the variables `yaxis`, `xaxis`, `factor`, `factor1` and `factor2`, and if we have created the object `shingles`, then the following formulae

```
yaxis~xaxis|factor
yaxis~xaxis|factor1*factor2
yaxis~xaxis|shingles*factor
```

can be used to plot the points for which the coordinates are contained within `xaxis` and `yaxis`. This will condition, according to a) the categories of

factor, b) the confrontation of the categories of `factor1`, `factor2` and c)
those of `factor` and the intervals defined in `shingles`.

Below we present two examples from the `states.x77` table available in R.
We take the table `states.x77` and add the names of the different states and
regions of the United States:

```
> data(state)
> states <- data.frame(state.x77,state.name=rownames(state.x77),
    state.region = state.region)
```

We examine the representation of crimes according to population in the four
regions of the United States (simple conditioning, Figure 3.28), and then add
the variable income (multiple conditioning). To conduct this second conditioning, income must be broken down; we here choose three classes:

```
> income <- equal.count(states[,"Income"],number=3)
```

We draw two graphs:

```
> xyplot(Murder ~ Population | state.region, data = states)
```

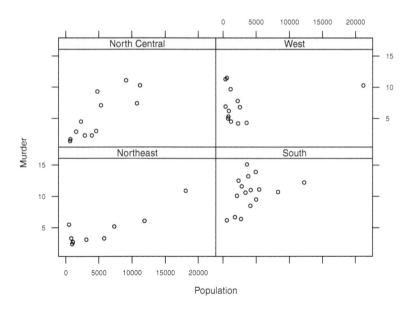

Figure 3.28
Crime rates and population size by region.

```
> xyplot(Murder ~ Population | state.region*income,
    data = states)
```

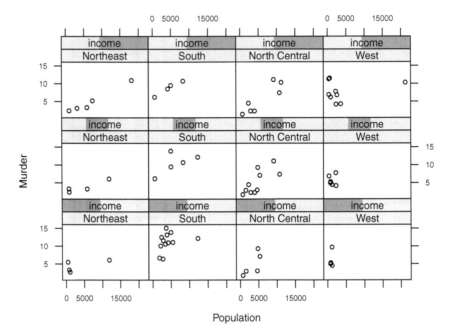

Figure 3.29
Crime rates and population size by region and class of income.

We thus obtain 4 × 3 panels in which we can visualise the relationship between crime rate and population for each class of income × region (Figure 3.29). This graph features numerous panels. It can often be easier to read if one of the variables is represented using a specific colour or a symbol. For each class of income we represent the crime rate with different symbols / colours for each region. On the computer screen, the different colours are visible; however, if the graphs are exported in black and white, it is better to use symbols. Within each panel (each class of income), there are four groups (the regions).

```
> xyplot(Murder~Population|income, groups=state.region,
    data=states,auto.key=TRUE,type="p")
```

A legend can be added using the argument `auto.key`, which is used to control the appearance of the legends: visualisation of lines (instead of symbols), position (top, right, bottom, or left), etc. (Figure 3.30).

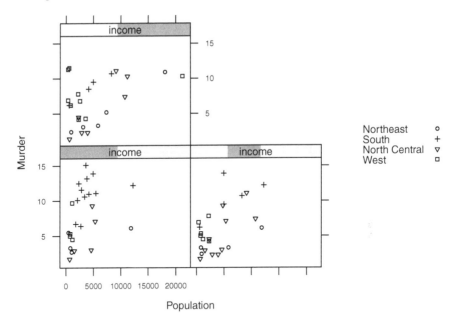

Figure 3.30
Crime rates and population size by class of income using postscript theme
(see Section 3.2.4, p. 81).

3.2.3   Customising Graphs

3.2.3.1   Panel Function

The `panel` option is used to more accurately control the appearance of the
graph within each panel. Each "lattice" type graph function (**bwplot**, **xyplot**,
etc., see Table 3.2 for other examples) has its own panel function (**panel.bwplot**,
panel.xyplot, etc.).   Other panel functions are also available, such as loess
smoothing (**panel.loess**), linear regression, etc. (the most useful panel functions can be obtained with `help(panel.functions)`). Of course, it is possible
to create one's own panel functions, but the graph commands must be compatible with the lattice package: for example, to insert text, one must not use
text, but **ltext** (l for lattice). For further information, we recommend the book
by Murrel (2005). However, it is relatively simple to merge two panel functions together. In the following example (see Figure 3.31), we shall smooth
the crime rate according to the population sizes (**panel.loess**) and draw the
points which correspond to the observations (**panel.xyplot**):

```
> mypanel <- function(...) {
+     panel.loess(...)
+     panel.xyplot(...)
+ }
```

```
> xyplot(Murder ~ Population | income,data=states,panel=mypanel)
```

The **mypanel** function may seem strange, but the ... are indeed part of the function, and are not arguments to be detailed.

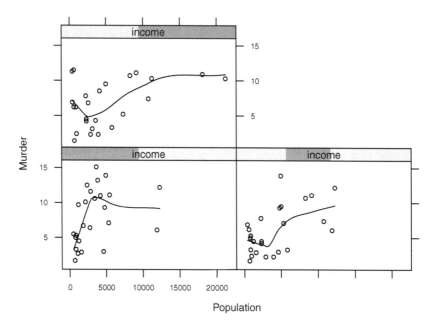

Figure 3.31
Crime rates according to population size by class of income: smoothing and data representation.

3.2.3.2 Controlling Size

Often, the characters are too large (or small), or the banner is too wide, and it can therefore be useful to know how to modify the size of these elements. The following command can be used to change the overall size of the symbols and text:

```
> fontsize <-trellis.par.get("fontsize")
> fonsize$text <- 8
> fontsize$points <- 5
> trellis.par.set("fontsize",fonsize)
```

The `fontsize` option is the name of the list, the values of which are within another list which controls the size of the points and the font. In order to control the size of the points in relation to the default size (option `cex`), another option is also available:

```
> plot.symbol <- trellis.par.get("plot.symbol")
```

```
> plot.symbol$cex <- 0.5
> trellis.par.get("plot.symbol",plot.symbol)
```

For a complete list of parameters, use **trellis.par.get()**.

However, it is also possible to control symbol size, legend font size, and graduations (x and y axes) independently, using the following options:

```
> cex <- 0.5 # symbol magnification
> xlab <- list("my title for x-axis",cex=0.7) # title for x axis
> ylab <- list(cex=0.7) # no title for y axis
> scales <- list(cex=0.5) # size of the graduations
```

3.2.3.3   Panel Layout

It is very simple to change the layout of the panels in a "lattice" graph. Simply use the `layout` argument from the graph functions. Thus, if we want to draw density estimators (**densityplot**, see Table 3.2) for the crime rate depending on class of income on a single row, simply request a layout with three panels on one row (Figure 3.32):

```
> densityplot(~Murder|income,data=states,layout=c(3,1))
```

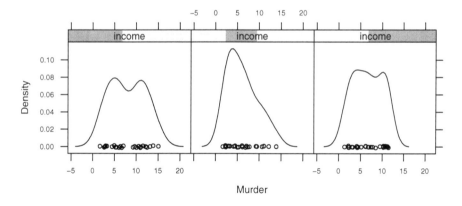

Figure 3.32

Crime rates according to population size by class of income: smoothing and data representation.

3.2.4   Exportation

To export "lattice" graphs, we use the same procedure as for conventional graphs (see Section 3.1.5, p. 66). It is sometimes necessary to export a graph in black and white. In such cases, we recommend that you replace the colours with different symbols. This tedious step is actually managed automatically using the **trellis.device** function. This function is used to manage the themes.

Themes control the appearance of the graph by suggesting a combination of colours, symbols and different types of lines. One simple way of visualising these themes is to use the **show.settings** function:

```
> show.settings(standard.theme("postscript")) # black and white
> show.settings(col.whitebg()) # default
> show.settings(standard.theme("quartz")) # quartz
```

To export a black and white "lattice" graph in pdf format, simply use

```
> trellis.device(device="pdf",theme = standard.theme("postscript"),
      file="export.pdf")
> my lattice expression(s)
> dev.off()
```

3.2.5   Other Packages

Other packages are available which can be used to create more advanced graphs. Here, we simply mention their existence. The rgl package is used to represent 3D data and to rotate them using the mouse. The ggplot2 package attempts to combine the positive elements of conventional and lattice graphs. Finally, the iplots package is used to create interactive graphs.

3.3   Exercises

Exercise 3.1 (Draw a function)
1. Draw the sine function between 0 and 2π (use **pi**).

2. Add the following title (**title**): plot of sine function.

Exercise 3.2 (Comparison of Distributions)
1. Draw the pdf (probability distribution function) of the standard Gaussian distribution between -4 and 4 (use **dnorm**).

2. On the same graph, draw Student's t-distribution to 5 and 30 degrees of freedom. Use the **curve** function and a different colour for each line.

3. Add a legend at the top left to differentiate between each distribution.

Exercise 3.3 (Plotting Points)
1. Import the ozone dataset and plot the scatterplot for maximum ozone (max03) and temperature (T12).

2. Plot the scatterplot max03 according to T12, with lines connecting the points.

3. Draw graph in Figure 3.33 using the **order** function.

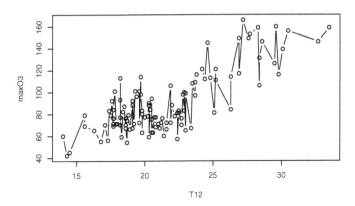

Figure 3.33
Scatterplot max03 according to T12.

Exercise 3.4 (Law of Large Numbers)

1. Having set the seed of the random generator (**set.seed**), simulate a sample $(x_1, ..., x_{1000})$ from Bernoulli's distribution with parameter $p = 0.6$.

2. Calculate the successive means $M_l = S_l/l$ where $S_l = \sum_{i=1}^{l} X_i$. Draw M_l according to l, then add the horizontal line with equation $y = 0.6$.

Exercise 3.5 (Central Limit Theorem)

1. Let us denote X_1, X_2, ... X_N i.i.d. random variables following Bernoulli's distribution with parameter p. Recall the distribution of $S_N = X_1 + ... + X_N$. Specify the mean and standard deviation.

2. Set $p = 0.5$. For $N = 10$, using the **rbinom** function, simulate $n = 1000$ occurrences S_1, ..., S_{1000} of a binomial distribution with parameters N and p. Organise the quantities $\frac{S_i - N \times p}{\sqrt{N \times p \times (1-p)}}$ into a vector U10. Do the same with $N = 30$ and $N = 1000$ to obtain two new vectors U30 and U1000.

3. In one window (**par(mfrow)**) represent histograms for U10, U30 and U1000, each time overlapping the density of the standard Gaussian distribution (obtained using **dnorm**).

Exercise 3.6 (Draw Sunspots)

1. Read data from file sunspots.csv series which details, date by date, the relative number of sunspots. Check the variable types after reading.

2. Create a qualitative variable **thirty** with a value of 1 for the first year (1749), increasing by 1 every thirty years. In order to do so, round up the result (**floor**) of the division by 30 (or the division integer).

3. Enter the `colour` vector which contains the following colours: green, yellow, magenta, orange, cyan, grey, red, green and blue. Automatically check that these colours are indeed contained within the **colors()** vector (instructions **%in%** and **all**).

4. Draw the time series as seen in Figure 3.34. Use the **palette** function, **plot**, **lines** and a loop (see also **unique**).

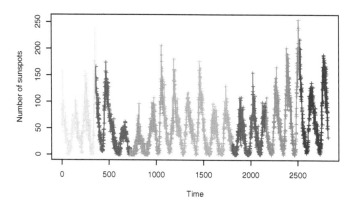

Figure 3.34
Sunspots according to observation number.

Exercise 3.7 (Draw a Density)
The objective of this exercise is to reproduce the following graph (Figure 3.35).

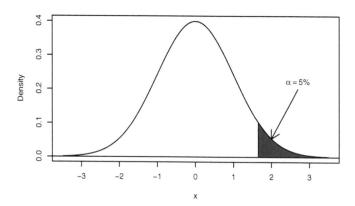

Figure 3.35
Density of a normal distribution.

1. Draw the density of the random variable X, $X \sim \mathcal{N}(0,1)$ (see **dnorm**).

2. Add the abscissa axis (see **abline**).

3. Identify the area below the curve to the right of q corresponding to a probability of 5% and fill it blue (**polygon**); see Figure 3.35.

4. Add an arrow identifying the coloured area (**arrows**).

5. At the end of the arrow, write $\alpha = 5\%$ (**text** and **expression**).

Exercise 3.8 (Multiple Graphs)
1. Reproduce the graph in Figure 3.36.

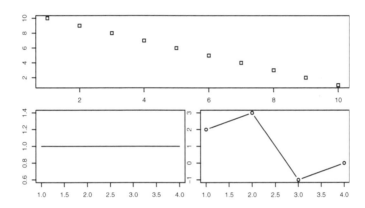

Figure 3.36
Three graphs on two rows.

2. Reproduce the same graph as above but with a graph at the bottom right with a width of 1 for 4 compared with that at the bottom left (see the arguments for **layout**).

Exercise 3.9 (Draw Points and Lines)
1. Import the ozone table.

2. Use the lattice package to draw the scatterplot max03 according to T12.

3. Use the lattice package to draw the scatterplot max03 according to T12 with a line connecting the points (type "p" and type "l").

4. Use the lattice package to draw the scatterplot max03 according to T12, with lines connecting the points (type "a"). Compare this line to that in Figure 3.33 (Exercise 3.3, p. 83). What changed using the type="a" option specific to the lattice package?

Exercise 3.10 (Panel)
1. Construct a data-frame with the first three rows of max03 and T12 from the ozone table.

2. Concatenate the table according to the rows so that the table appears six times.

3. Construct an ordinal factor for which the observations are 3 times `"h"`, 3 times `"b"`, 3 times `"p"`, 3 times `"l"`, 3 times `"s"` and 3 times `"S"`. The order of levels will be `"p"`, `"l"`, `"b"`, `"h"`, `"s"` and `"S"`.

4. Add this ordinal factor to the data of question 2.

5. Reproduce graph in Figure 3.10 (p. 58) with **xyplot** and the appropriate function panel (see `subscripts` in the help section of **xyplot**).

6. Reproduce the same graph but instead of using an ordinal factor, use a non-ordinal (classical) factor and observe the difference.

4

Making Programs with R

Programming in R is based on the same principles as for other software for scientific calculations. It therefore uses familiar programming structures (loops, condition `if else`, etc.) as well as predefined functions specific to statistical practices.

4.1 Control Flows

4.1.1 Grouped Expressions

A group of commands is similar to brackets in mathematics: Those commands which are grouped together are executed together. In R, groups of commands are contained within curly brackets:

```
> {
+   expr1
+   expr2
+   ...
+ }
```

Two successive commands are separated by a new line (Enter key). However, it is possible to separate two commands and to keep them on the same line. To do this, simply separate them with a semi-colon. Therefore the above example could also be written as

```
> { expr1; expr2; ... }
```

4.1.2 Loops (for or while)

Conventional loops are available in R. Let us start with the `for` loop. We want to display all the integers from 1 to 99. One solution is as follows:

```
> for (i in 1:99) print(i)
```

The i index takes as its values all the coordinates of the chosen vector. If we want to display only odd integers, simply create a vector which starts at 1 and goes up to 99, at intervals of two. The **seq** function is used to create just such a vector. The loop becomes

```
> for (i in seq(1,99,by=2)) print(i)
```

This method can easily be generalised to any vector. Thus, if we choose a character vector which represents the first three days of the week, we obtain

```
> vector <- c("Monday","Tuesday","Wednesday")
> for (i in vector) print(i)
[1] "Monday"
[1] "Tuesday"
[1] "Wednesday"
```

In general there are multiple orders to be executed for each iteration. To do so, the commands must be grouped together. Generally speaking, the `for` loop is expressed:

```
> for (i in vector) {
+    expr1
+    expr2
+    ...
+ }
```

Another possibility of a loop is the `while` condition. Its general syntax is as follows:

```
> while (condition) {
+    expr1
+    expr2
+    ...
+ }
```

The orders `expr1`, `expr2`, etc. are executed as long as the condition is true, and this is evaluated at the beginning of the loop. As soon as the condition is found to be false, the loop is stopped. Thus,

```
> i <- 1
> while (i<3) {
+    print(i)
+    i <- i+1
+ }
[1] 1
[1] 2
```

is used to display i and to increase it by 1 when i is less than 3.

One final possibility for a loop is the order `repeat`. It is understood as: repeat the orders indefinitely. To ensure the loop is stopped, we use the order `break`. This order can be used for any loop. An example is given in the following subsection.

4.1.3 Conditions (if, else)

This involves executing an order under condition: the order is executed if, and only if, the condition is true. In its simplest form it is written as:

```
> if (condition) {
+    expr1
+    expr2
+    ...
+ }
```

For example, if we want to use a **repeat** loop to print i varying from 1 to 3 inclusive, we must come out of the loop prior to printing, and when i is greater than 3,

```
> i <- 1
> repeat {
+    print(i)
+    i <- i+1
+    if (i>3) break
+ }
```

Here, if i is greater than 3, we only need to execute a (**break**) order. Therefore the commands do not need to be grouped together (the curly brackets can be omitted).

Another condition can be added after **if**: the **else** condition, which can be used to distinguish between two cases. Either the condition is true, in which case the order (or group of orders) after **if** is executed, or it is not true and therefore the order (or group of orders) after **else** is executed. In its most general form, the **if, else** condition is written as:

```
> if (condition) {
+    expr1
+    expr2
+    ...
+ } else {
+    expr3
+    expr4
+    ...
+ }
```

Be aware that the order **else** must be on the same line as the "}" closing bracket of the **if** clause (as above).

4.2   Predefined Functions

Certain functions in R are predefined in order to avoid having to use loops which generally have lengthy calculation times. The most common of these functions is undoubtedly **apply**, which is used to apply one function to all the margins of a table. Let us consider the table X made up of twenty randomly chosen integers between 1 and 20 (**sample**). The mean for each column is calculated as follows:

```
> set.seed(1234)
> X <- matrix(sample(1:20,20),ncol=4)
> X
     [,1] [,2] [,3] [,4]
[1,]    3   10    7    9
[2,]   12    1    5   17
[3,]   11    4   20   16
[4,]   18    8   15   19
[5,]   14    6    2   13
> apply(X,MARGIN=2,FUN=mean)
[1] 11.6  5.8  9.8 14.8
```

It is also possible to add additional arguments to the function which is applied to each column. For example, if there is a missing piece of data in X, it can be interesting to use the argument na.rm=TRUE of the **mean** function, which only calculates the mean of the present data:

```
> X[1,1] <- NA
> apply(X,MARGIN=2,FUN=mean)
[1]    NA  5.8  9.8 14.8
> apply(X,MARGIN=2,FUN=mean,na.rm=TRUE)
[1] 13.75  5.80  9.80 14.80
```

Due to the frequent use in statistics of means per column (or row), there is a shortcut in the form of a function **colMeans** (or **rowMeans**):

```
> colMeans(X,na.rm=TRUE)
[1] 13.75  5.80  9.80 14.80
```

In the same way, column (or row) sums can also be executed directly using **colSums** (or **rowSums**).

   If the data table is three dimensional (data cube), it is possible to execute the function by row (MARGIN=1), by column (MARGIN=2) or by depth (MARGIN=3). It is also possible to execute the function for confronting rows and columns (MARGIN=c(1,2)), or rows and depth (MARGIN=c(1,3)) or columns and depth (MARGIN=c(2,3)). As an example, let us calculate the sum of a table with three entries for each row/column pair:

```
> set.seed(1234)
> Y <- array(sample(24),dim=c(4,3,2))
> Y
, , 1
     [,1] [,2] [,3]
[1,]    3   18   11
[2,]   15   13    8
[3,]   14    1   10
[4,]   22    4   23

, , 2
     [,1] [,2] [,3]
[1,]   17   21    5
[2,]   16    2   20
[3,]   24    7    9
[4,]   19    6   12
> apply(Y,MARGIN=c(1,2),FUN=sum,na.rm=TRUE)
     [,1] [,2] [,3]
[1,]   20   39   16
[2,]   31   15   28
[3,]   38    8   19
[4,]   41   10   35
```

Here we have used the R functions **mean** and **sum**, but we could just as easily have used functions which we programmed in advance. For example,

```
> MyFunction <- function(x,y) {
+    z <- x+sqrt(y)
+    return(1/z)
+ }
> set.seed(1234)
> X <- matrix(sample(20),ncol=4)
> X
     [,1] [,2] [,3] [,4]
[1,]    3   10    7    9
[2,]   12    1    5   17
[3,]   11    4   20   16
[4,]   18    8   15   19
[5,]   14    6    2   13
> apply(X,MARGIN=c(1,2),FUN=MyFunction,y=2)
            [,1]       [,2]       [,3]       [,4]
[1,] 0.22654092 0.08761007 0.11884652 0.09602261
[2,] 0.07454779 0.41421356 0.15590376 0.05430588
[3,] 0.08055283 0.18469903 0.04669796 0.05742436
[4,] 0.05150865 0.10622236 0.06092281 0.04898548
[5,] 0.06487519 0.13487607 0.29289322 0.06937597
```

Many functions are based on the same principle as the **apply** function. For example, the **tapply** function applies one same function, not to the margins of the table but instead at each level of a factor or combination of factors:

```
> Z <- 1:5
> Z
[1] 1 2 3 4 5
> vec1 <- c(rep("A1",2),rep("A2",2),rep("A3",1))
> vec1
[1] "A1" "A1" "A2" "A2" "A3"
> vec2 <- c(rep("B1",3),rep("B2",2))
> vec2
[1] "B1" "B1" "B1" "B2" "B2"
> tapply(Z,vec1,sum)
A1 A2 A3
 3  7  5
> tapply(Z,list(vec1,vec2),sum)
   B1 B2
A1  3 NA
A2  3  4
A3 NA  5
```

lapply, or its equivalent, **sapply**, applies one same function to every element in a list. The difference between these two functions is as follows: The **lapply** function by default yields a list when the **sapply** function yields a matrix or a vector. Let us therefore create a list containing two matrices and then calculate the mean of each element of the list; here is each matrix:

```
> set.seed(545)
> mat1 <- matrix(sample(12),ncol=4)
> mat1
     [,1] [,2] [,3] [,4]
[1,]    9    4    1   11
[2,]   10    2    3    6
[3,]    7    5    8   12
> mat2 <- matrix(sample(4),ncol=2)
> mat2
     [,1] [,2]
[1,]    4    3
[2,]    2    1
> mylist <- list(matrix1=mat1,matrix2=mat2)
> lapply(mylist,mean)
$matrix1
[1] 6.5

$matrix2
[1] 2.5
```

It is even possible to calculate the sum by column of each element of the list using the **apply** functions as FUN function of the **lapply** function:

```
> lapply(mylist,apply,2,sum,na.rm=T)
$matrix1
[1] 26 11 12 29

$matrix2
[1] 6 4
```

The **aggregate** function works on data-frames. It separates the data into sub-groups defined by a vector, and calculates a statistic for all the variables of the data-frame for each sub-group. Let us reexamine the data which we generated earlier and create a data-frame with two variables Z and T:

```
> Z <- 1:5
> T <- 5:1
> vec1 <- c(rep("A1",2),rep("A2",2),rep("A3",1))
> vec2 <- c(rep("B1",3),rep("B2",2))
> df <- data.frame(Z,T,vec1,vec2)
> df
  Z T vec1 vec2
1 1 5   A1   B1
2 2 4   A1   B1
3 3 3   A2   B1
4 4 2   A2   B2
5 5 1   A3   B2
> aggregate(df[,1:2],list(FactorA=vec1),sum)
  FactorA Z T
1      A1 3 9
2      A2 7 5
3      A3 5 1
```

Sub-groups can also be defined by a vector generated by two factors:

```
> aggregate(df[,1:2],list(FactorA=vec1,FactorB=vec2),sum)
  FactorA  FactorB Z T
1      A1       B1 3 9
2      A2       B1 3 3
3      A2       B2 4 2
4      A3       B2 5 1
```

The **sweep** function is used to apply one single procedure to all the margins in a table. For example, if we want to centre and standardise the columns of a matrix X, we write

```
> set.seed(1234)
> X <- matrix(sample(12),ncol=3)
```

```
> X
     [,1] [,2] [,3]
[1,]    2   10    3
[2,]    7    5    8
[3,]   11    1    4
[4,]    6   12    9
> mean.X <- apply(X,2,mean)
> mean.X
[1] 6.5 7.0 6.0
> sd.X <- apply(X,2,sd)
> sd.X
[1] 3.696846 4.966555 2.943920
> Xc <- sweep(X,2,mean.X,FUN="-")
> Xc
      [,1] [,2] [,3]
[1,] -4.5    3   -3
[2,]  0.5   -2    2
[3,]  4.5   -6   -2
[4,] -0.5    5    3
> Xcr <- sweep(Xc,2,sd.X,FUN="/")
> Xcr
           [,1]        [,2]        [,3]
[1,] -1.2172540  0.6040404 -1.0190493
[2,]  0.1352504 -0.4026936  0.6793662
[3,]  1.2172540 -1.2080809 -0.6793662
[4,] -0.1352504  1.0067341  1.0190493
```

It must be noted that to centre and standardise table X, we can more simply use the **scale** function. The **by** function, however, is used to apply one function to a whole data-frame for the different levels of a factor or list of factors. This function is thus similar to the **tapply** function, as it is used with a data-frame rather than a vector. Let us generate some data:

```
> set.seed(1234)
> T <- rnorm(100)
> Z <- rnorm(100)+3*T+5
> vec1 <- c(rep("A1",25),rep("A2",25),rep("A3",50))
> don <- data.frame(Z,T)
```

We can thus obtain a summary of each variable for each level of the factor vec1:

```
> by(don,list(FactorA=vec1),summary)
FactorA: A1
       Z                   T
 Min.   :-2,540    Min.    :-2.3457
 1st Qu.: 2.380    1st Qu.:-0.7763
```

```
Median : 3.662    Median :-0.4907
Mean   : 4.331    Mean   :-0.2418
3rd Qu.: 5.737    3rd Qu.: 0.2774
Max.   :12.221    Max.   : 2.4158
------------------------------------
FactorA: A2
        Z                   T
 Min.   :-1.856    Min.   :-2.1800
 1st Qu.: 1.520    1st Qu.:-1.1073
 Median : 3.071    Median :-0.8554
 Mean   : 2.991    Mean   :-0.6643
 3rd Qu.: 4.050    3rd Qu.:-0.4659
 Max.   : 9.321    Max.   : 1.4495
------------------------------------
FactorA: A3
        Z                   T
 Min.   :-0.7953   Min.   :-1.80603
 1st Qu.: 3.2903   1st Qu.:-0.56045
 Median : 5.1260   Median :-0.04396
 Mean   : 5.4811   Mean   : 0.13953
 3rd Qu.: 8.0739   3rd Qu.: 0.81208
 Max.   :12.6221   Max.   : 2.54899
```

Be careful: If we request the sum, we obtain the sum for all the variables for each level of the factor vec1 rather than the sum variable by variable for each level of vec1:

```
> by(don,list(FactorA=vec1),sum)
FactorA: A1
[1] 102.2201
-------------
FactorA: A2
[1] 58.16823
-------------
FactorA: A3
[1] 281.0313
```

It is also possible to perform more complex calculations. For example, if we want to calculate a regression for each level of the variable vec1, we can call on **by** to repeat a function which we will define ourselves (Section 4.3), which here is the regression of variable Z according to variable T from the dataset x. This function's only argument is x but it also calculates the coefficients of linear regression.

```
> myfunction <- function(x){
+    summary(lm(Z~T, data=x))$coef
+ }
```

```
> by(don, vec1, myfunction)
vec1: A1
              Estimate Std. Error   t value      Pr(>|t|)
(Intercept) 5.066900   0.1732763 29.24174 1.073704e-19
T           3.045363   0.1850066 16.46084 3.214797e-14
---------------------------------------------------------
vec1: A2
              Estimate Std. Error   t value      Pr(>|t|)
(Intercept) 5.090834   0.2419277 21.04279 1.592367e-16
T           3.160780   0.2340362 13.50551 2.018228e-12
---------------------------------------------------------
vec1: A3
              Estimate Std. Error   t value      Pr(>|t|)
(Intercept) 5.081208   0.1700007 29.88935 1.120636e-32
T           2.865977   0.1640595 17.46914 1.939013e-22
```

This type of calculation is also automated in a different way by the **lmList** function available in the nlme package.

There are also other functions such as the **replicate** function, which is used to repeat an expression **n** times (see Table 4.1). Usually, the expression includes a random number generation, and evaluating this expression **n** times does not yield the same result. This therefore means making a loop for (i in 1:n). For example,

```
> set.seed(1234)
> replicate(n=8, mean(rnorm(100)))
[1] -0.224412908  0.165109001  0.129179613 -0.012143006
[5]  0.055836342 -0.002691679  0.091426757  0.078374289
```

The **outer** function is used to repeat a function for each occurrence of the combination of two vectors. For example,

```
> Month <- c("Jan","Feb","Mar")
> Year <- 2008:2010
> outer(Month,Year,FUN="paste")
     [,1]         [,2]         [,3]
[1,] "Jan 2008"  "Jan 2009"  "Jan 2010"
[2,] "Feb 2008"  "Feb 2009"  "Feb 2010"
[3,] "Mar 2008"  "Mar 2009"  "Mar 2010"
```

Again, it is possible here to add arguments to the chosen function. Here, we separate the months and the years by a dash rather than a space, which is the default character:

```
> outer(Month,Year,FUN="paste",sep="-")
     [,1]         [,2]         [,3]
[1,] "Jan-2008"  "Jan-2009"  "Jan-2010"
[2,] "Feb-2008"  "Feb-2009"  "Feb-2010"
[3,] "Mar-2008"  "Mar-2009"  "Mar-2010"
```

TABLE 4.1
Summary of a Few Functions that Can Be Used to Avoid loops

Function	Description
aggregate	Applies a function to a subset of rows from a data-frame (mean of all the variables by sex, etc.). Yields a data-frame (usually)
apply	Applies a function to a margin (row, column, etc.) of a matrix or table with multiple entries (column sum, row sum, etc.)
by, **tapply**	Applies a function to a subset of rows of a data-frame or vector (mean of all variables by sex, etc.). Generally yields a list
mapply	Repeats the application of a function taking the first coordinates of each vector as its first argument, the second coordinates of each vector as its second argument, etc. (see also **Vectorize**)
outer	Applies a (vectorised) function to each pair of two vectors
replicate	Evaluates the same expression n times
sapply, **lapply**	Applies a function to each coordinate of a vector or each component of a list. Yields a vector, a matrix or a list
sweep	Applies a function to a margin (row, column, etc.) of a matrix or table with multiple entries with a `STATS` argument which varies with each margin (divides each column by its standard deviation, substracts the mean from each column, etc.)

4.3   Creating a Function

Functions are used to carry out many different R commands. These orders can depend on arguments provided as input, but this is not always the case. Functions provide unique result objects (see Figure 4.1). These resulting objects are specified within the functions by the **return** function. By default, if the written function yields no results, the last result obtained prior to output is given as the result.

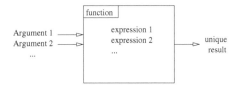

Figure 4.1
Diagram of an R function.

Let us begin with a simple example: the sum of the first n integers. The number n is an integer which is the input argument; the result is simply the sum, as requested:

```
> mysum <- function(n) {
    result <- sum(1:n)
    return(result)
}
```

The function is called upon using its name (**mysum**), followed by the input arguments between brackets. Here, the function has an argument as an input. It is therefore called upon by simply using

```
> mysum(3)
[1] 6
```

It is possible to attribute the result of a function to an R object using

```
> res <- mysum(3)
> res
[1] 6
```

This function can be improved as follows:

```
> mysum <- function(n) {
+    if (n<=0) stop("n must be a positive integer")
+    if (floor(n)!=n) warning(paste("rounds",n,"as",floor(n)))
+    result <- sum(1:floor(n))
+    return(result)
+ }
```

Of course, when n is not larger than one, the sum is not meaningful. The function is thus stopped (**stop**) and displays an error message. On the other hand, when n is larger than one but not a whole number, the function rounds n down to its integer part (**floor**) and sends a message indicating that n has been replaced (**warnings**).

The function is called in much the same way:

```
> mysum(4.325)
[1] 10
Warning message:
In mysum(4.325) : rounds 4.325 as 4
```

We now suggest a function with two input arguments: `factor1` and `factor2`, two qualitative variables. This function yields the contingency table as well as the character vector for the levels of `factor1` and `factor2`, for which the combined sample size is zero. Here, more than one result will be given. As one is a table and the other a character vector (or character matrix), these two objects can neither be brought together in a matrix (as they are not of the same type), nor can they be combined in a data-frame (as they are not the same length). The only result will therefore be a list grouping together these two results.

The function will therefore calculate the contingency table (**table**) and then select the void cells. We then need to know the indices corresponding to the void cells of the contingency table (**which** function, option `arr.ind=TRUE`) and identify the names of the corresponding categories:

```
> myfun <- function(factor1,factor2) {
+    res1 <- table(factor1,factor2)
+    selection <- which(res1==0,arr.ind = TRUE)
+    res2 <- matrix("",nrow=nrow(selection),ncol=2)
+    res2[,1] <- levels(factor1)[selection[,1]]
+    res2[,2] <- levels(factor2)[selection[,2]]
+    return(list(tab=res1,level=res2))
+ }
```

If we call upon the function with the factors `wool` and `tension`:

```
> tension <- factor(c(rep("Low",5),rep("High",5)))
> wool <- factor(c(rep("Mer",3),rep("Ang",3),rep("Tex",4)))
```

This yields

```
> myfun(tension,wool)
$tab
        factor2
factor1 Ang Mer Tex
   High   1   0   4
   Low    2   3   0

$level
     [,1]   [,2]
[1,] "High" "Mer"
[2,] "Low"  "Tex"
```

To find out more about using a function, see Section 1.5.

4.4   Exercises

Exercise 4.1 (Factorial)
1. Programme the factorial function: $n! = n \times (n-1) \times \cdots \times 2 \times 1$ using the **prod** function.

2. Programme the factorial function using a loop `for`.

Exercise 4.2 (Ventilation (allocation at random))
1. Repeat Exercise 2.6 (p. 49) and programme a ventilation function. As its arguments, this function will take both a variable and the threshold from which categories must be ventilated. Set the threshold to 5% by default.

2. Write a function so that all the qualitative variables in a table can be ventilated (see previous question).

Exercise 4.3 (Ventilation on an Ordered Factor)
1. Repeat Exercise 2.7 (p. 49) and programme a ventilation function on an ordered factor. As its arguments, this function will take both a variable and the threshold from which categories must be ventilated. Set the threshold to 5% by default.

2. Write a function so that all the ordered qualitative variables in a table can be ventilated (see previous question).

Part II

Statistical Methods

Introduction to the Statistical Methods Part

In this part we present most of the classical statistical methods in the form of worked examples. Each method is illustrated using a concrete example.

The structure of the worked examples is always the same. We start with a quick presentation of the method with theoretical reminders (section objective). Then follows a description of the dataset and the associated problem (section example). We present the steps to solve this problem (section steps). The example is processed step by step, and a brief commentary of the results is given (section processing the example). In this part, the different instructions are given using command lines. However, wherever possible, we use the R Commander interface available in the Rcmdr package (see Appendix A.3, p. 267) which allows one to perform the various methods using easy-to-use drop-down menus. In conclusion, a section called "Taking Things Further" gives reference books or less-standard extensions to the method.

Finally, it must be noted that, as each worked example is designed to be consulted independently of the others, there will certainly be some overlaps.

From a purely practical standpoint, each dataset analysed in each worked example is available on the website
http://www.agrocampus-ouest.fr/math/RforStat
There are therefore two ways of importing data, both of which must be done at the beginning of the study:

- Either by saving the file in the working directory and using the following command line

```
> mydata <- read.table("MyDataFile.csv", header=TRUE, sep=";")
```

- Or by reading the dataset directly from the website:

```
> mydata <- read.table("http://www.agrocampus-ouest.fr/math/
    RforStat/MyDataFile.csv", header=TRUE, sep=";")
```

We believe that learning R requires a solid knowledge base detailed in the first four chapters. However, teaching experience has shown that impatient users often try to learn to use the software on their own. For such users, here follows a brief summary of the previous sections.

5

A Quick Start with R

5.1 Installing R

Simply go to the CRAN website at `http://cran.r-project.org/` and install the version of R best suited to your computer's operating system. For comprehensive documentation on the installation process, visit `http://cran.r-project.org/doc/manuals`.

5.2 Opening and Closing R

In Windows and Mac OS X, look for R in Program Files. In Linux, simply type R in a command window. To close R, either use the menu or type the command **q()**. When exiting the program, the software will ask if you want to save the work session. This means that the next time that you open R, you will be able to retrieve all the objects you have constructed. Most of the time it is preferable to save the instructions rather than all the objects. To do so, write and save the instructions in a text file, known as a script.

5.3 The Command Prompt

When R is launched, a window opens and waits for some instructions with a prompt >. R can be used immediately as a calculator, for example,

```
> 2+3.2
[1] 5.2
```

The [1] indicates that the first (and only) coordinate of the resulting vector is 5.2. When R is waiting for the next instruction, the command prompt becomes +. For example, if you type

```
> 1 -
```

R will wait for the second part of the subtraction and the command prompt will be +. To substract 2 from 1, type

```
+ 2
[1] -1
```

Generally, either a) or " has been forgotten. Simply suggest one (or more) bracket(s) or speech mark(s) to end the command.

5.4   Attribution, Objects, and Function

R makes calculations using functions (which will be written in bold):

```
> sqrt(2)
[1] 1.414214
```

Results can be attributed to more or less complex objects using <- or =. Thus we create vector x with integer values from 3 to 7:

```
> x <- 3:7
> x
[1] 3 4 5 6 7
```

The content can be displayed simply by typing the name of the object.

5.5   Selection

To construct the vector y with coordinates 2 and 4 of vector x, type

```
> y <- x[c(2,4)]
> y
[1] 4 6
```

Select columns 1 and 3 of a matrix m and rows 2 and 5 respectively using

```
> m <- matrix(1:15,ncol=3)    #creates the matrix
> m[,c(1,3)]                   #selects columns 1 and 3
> m[c(2,5),]                   #selects rows 2 and 5
```

In combination, rows 4 and 2 of columns 2 and 3 are obtained as follows:

```
> m[c(4,2),c(2,3)]
      [,1] [,2]
[1,]    9   14
[2,]    7   12
```

We can also assign objects of different kinds to lists:

```
> mylist <- list(vector=x,mat=m)
```

The component `mat` from the list `mylist` is selected using

```
> mylist$mat
```

5.6 Other

All commands can be stopped using the shortcut `Ctrl + c` (or the STOP icon). It is possible to reuse preceding commands using the up or down arrows.

Additionally, to obtain help for the **mean** function, type

```
> help(mean)
```

5.7 Rcmdr Package

The simplest way to get to grips with R is undoubtedly to use the Rcmdr package which combines R with an easy-to-use drop-down menu. The advantage of this package is that is also provides the lines of code which correspond to the analyses carried out: Users can therefore familiarise themselves with the programming by seeing which functions are used. To use this package, install it and then launch it using

```
> library(Rcmdr)
```

5.8 Importing (or Inputting) Data

Usually, the first thing to do is to input or import data. Data can be easily imported from a spreadsheet (such as Excel). Generally speaking, data files contain the names of the variables in the first row and sometimes identifiers for the individuals in the first column. We recommend that this file be saved in a format which is easy to export, such as `txt` or `csv`. The file `myfile.csv` contains the values for two variables `x` and `y`, with measurements taken from fifty individuals. Let us suppose that the file is located at the same destination as R. The command by which the file can be imported and placed within an object named `dat` is

```
> dat <- read.table("myfile.csv", sep=";", header=TRUE, dec=".")
```

if for example the column separator is ";", the name of the variables is present on the first row (header=TRUE) and the decimal separator is ".". Then check that the reading was successful by summarising the dataset

```
> summary(dat)
```

5.9 Graphs

To construct a histogram and then a boxplot for all the individuals of the second variable, we write

```
> hist(dat[,2])
> boxplot(dat[,2])
```

If the second variable is called y, we can use the name of the variable rather than selecting the column by writing dat[,"y"] instead of dat[,2]. If the first variable is called x, y is represented in confrontation with x by writing

```
> plot(y~x,data=dat)
```

If the variable x is quantitative, the graph is a scatterplot. If x is qualitative, the graph is a boxplot for each category of x.

5.10 Statistical Analysis

Different functions can be used to conduct various statistical analyses. The easiest way to know how to use a function is to refer to its help section by typing help(myfunction). The help section demonstrates how the function is defined, all of its arguments and, finally, usage examples that can be directly copied and pasted into R. If the outputs of the function are assigned to an object named result, it is possible to list all the objects present in result using names(result). For example, after the regression of y according to x,

```
> result <- lm(y~x,data=dat)
> names(result)
 [1] "coefficients"  "residuals"    "effects"    "rank"
 [5] "fitted.values" "assign"       "qr"         "df.residual"
 [9] "xlevels"       "call"         "terms"      "model"
```

The result$coefficients command gives access to the object containing the regression coefficients. For certain functions, the **summary** function can be used to summarise the main results.

6

Hypothesis Test

6.1 Confidence Intervals for a Mean

6.1.1 Objective

The aim of this worked example is to construct a confidence interval for the mean of a quantitative variable from a sample. The method presented here can be extended to construct a confidence interval of a proportion (see Section 6.1.6, "Taking Things Further").

The confidence interval for the population mean μ for a variable X is expressed as:

$$\left[\bar{x} - \frac{\hat{\sigma}}{\sqrt{n}} \times t_{1-\alpha/2}(n-1) \, , \, \bar{x} + \frac{\hat{\sigma}}{\sqrt{n}} \times t_{1-\alpha/2}(n-1) \right]$$

where $\bar{x}$ is the empirical mean of the sample, $\hat{\sigma}$ the estimated standard deviation, n the sample size, and $t_{1-\alpha/2}(n-1)$ the quantile $1-\alpha/2$ of the Student's t-distribution with $(n-1)$ degrees of freedom. It is important to remember that the procedure assumes that the estimator of the mean $\bar{X}$ follows a normal distribution. This is true if X follows a normal distribution or if the sample size is large enough (in practice, $n > 30$, thanks to the central limit theorem).

6.1.2 Example

We examine the weight of adult female octopuses. We have access to a sample from 240 female octopuses fished off the coast of Mauritania. For the overall population, we would like to obtain an estimation of the mean of the weight and a confidence interval for this mean with a threshold of 95%.

6.1.3 Steps

1. Read the data.

2. Calculate the descriptive statistics (mean, standard deviation).

3. Construct a histogram.

4. (Optional) Test the normality of the data.

5. Construct the confidence interval.

6.1.4   Processing the Example

1. Reading the data:

Import the data by using the **read.table** function. Then check that the data has been correctly imported:

```
> octopus <- read.table("OctopusF.txt",header=T)
> dim(octopus)
[1] 240    1
> summary(octopus)

      Weight
 Min.    :   40.0
 1st Qu.:  300.0
 Median :  545.0
 Mean   :  639.6
 3rd Qu.:  800.0
 Max.   : 2400.0
```

Check that the sample size is equal to $n = 240$. Then check that the variable `Weight` is indeed quantitative and obtain the minimum, the mean, the quartiles and the maximum for this variable.

2. Calculating the descriptive statistics:

Next, estimate the mean ($\bar{x}$) and the standard deviation ($\hat{\sigma}$) within the sample:

```
> mean(octopus)
  Weight
639.625
> sd(octopus)
   Weight
445.8965
```

3. Constructing a histogram:

Prior to any statistical analysis, it can be useful to represent the data graphically. As the data is quantitative, a histogram is used:

```
> hist(octopus[,"Weight"],main="",nclass=15,freq=FALSE,
     ylab="Density",xlab="Weight")
```

We use the argument `freq=FALSE`, which is equivalent to `prob=TRUE`, to obtain a representation of the density, rather than frequency. The argument `nclass` is not compulsory but can be used to choose the number of intervals. The histogram (Figure 6.1) gives an idea of the weight distribution.

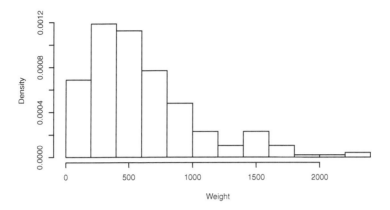

Figure 6.1
Histogram of octopus weight.

This distribution is asymmetric and is far from a normal distribution.

4. (Optional) Testing the normality of the data:

In order to construct a mean confidence interval, it is assumed that its estimator follows a normal distribution. For any large sample (over thirty individuals), this is a reasonable hypothesis. For small samples, the normality of the sample should be tested, rather than that of the mean. In order to do so, it is possible to have an account of probability plotting for informal assessment of normality (using the **qqnorm** function) or to use the Shapiro-Wilk test (**shapiro.test**). If normality is refused, we can construct a confidence interval using, for example, a bootstrap procedure (use the function **boot.ci** from the package **boot**). In this example, the sample size is 240 and these tests are therefore unnecessary.

5. Constructing the confidence interval:

The confidence interval is calculated using the function **t.test**:

```
> t.test(octopus$Weight,conf.level=0.95)$conf.int
[1] 582.9252 696.3248
attr(,"conf.level")
[1] 0.95
```

Strictly speaking, the unknown mean μ of X, estimated to be 639.625 g, either does or does not belong to the interval $[583, 696]$. It is therefore within the interval with a probability of 0 or 1. However, the procedure as a whole guarantees that, if it is repeated infinitely with new samples of the same size $n = 240$, then 95% of the confidence intervals will contain the true unknown value of μ. In practice, it is faster to say that the true mean μ is between 583 g and 696 g at a confidence level of 95%.

It is also possible to calculate the confidence interval "by hand" by calculating $\bar{x} \pm t_{239}(0.975) \times \hat{\sigma}/\sqrt{n}$, where $t_{239}(0.975)$ is the quantile 97.5% of the Student's t-distribution with 239 degrees of freedom:

```
> mean(octopus$Weight)-qt(0.975, df=239)*
      sd(octopus$Weight)/sqrt(240)
[1] 582.9252
> mean(octopus$Weight)+qt(0.975, df=239)*
      sd(octopus$Weight)/sqrt(240)
[1] 696.3248
```

6.1.5　Rcmdr Corner

1. Reading the data from a file:

`Data → Import data → from text file, clipboard, or URL ...`
then select the file.

To check that the dataset has been imported successfully:
`Statistics → Summaries → Active data set`

2. Calculating the descriptive statistics:

`Statistics → Summaries → Numerical summaries...`

3. Constructing a histogram:

`Graphs → Histogram...`　　　Then choose the number of classes and the scale of the `Densities` axes.

4. (Optional) Testing the normality of the data: `Statistics → Summaries → Shapiro-Wilk test of normality...`

5. Constructing the confidence interval:

We use a function which tests the equality of the mean with a predefined value (which is 0 by default) and as output this function provides the confidence interval: `Statistics → Means → Single-sample t-test...`

6.1.6　Taking Things Further

To construct a confidence interval for a proportion, we use the function **prop.test** (see also the worked example on test for equality of two proportions in Section 6.4, p. 125).

　　These classical methods are explained in detail in many books such as Clarke and Cooke (2004).

6.2 Chi-Square Test of Independence

6.2.1 Objective

The objective of this worked example is to test the independence between two qualitative variables. First, we consider that the data is in a contingency table (two-way table). In Section 6.2.6, "Taking Things Further", we examine a case in which the data is in a table of individuals × variables.

To test the independence of two qualitative variables, we test the null hypothesis H_0: "the two variables are independent" against the alternative hypothesis H_1: "the two variables are not independent". In order to do so, we calculate the following test statistic:

$$\chi^2 = \sum_{i=1}^{I} \sum_{j=1}^{J} \frac{(n_{ij} - T_{ij})^2}{T_{ij}},$$

where n_{ij} is the number of individuals (observed frequency) who take the category i of the first variable and the category j of the second, and T_{ij} corresponds to the expected frequency under the null hypothesis, and I and J are the number of categories for each of the variables. Thus, $T_{ij} = n\hat{p}_{i\bullet}\hat{p}_{\bullet j}$ with n the total sample size, $\hat{p}_{i\bullet} = \frac{\sum_{j} n_{ij}}{n}$ and $\hat{p}_{\bullet j} = \frac{\sum_{i} n_{ij}}{n}$. Under H_0, χ^2 follows a chi-square distribution with $(I-1) \times (J-1)$ degrees of freedom.

If there is no independence, it is interesting to calculate the contribution of pairs of categories to the chi-square statistic in order to observe associations between categories.

6.2.2 Example

This is a historic example by Fisher. He studied the hair colour of boys and girls in a Scottish county:

	Fair	Red	Medium	Dark	Jet Black
Boys	592	119	849	504	36
Girls	544	97	677	451	14

We would like to know if hair colour is independent of sex with a type-one error rate of 5%. If hair colour does depend on sex, we would like to study the association between categories, for example to know which colours are more common in girls than in boys.

6.2.3 Steps

1. Input the data.

2. Visualise the data.

3. (Optional) Calculate row and column profiles.

4. Construct the chi-square test.

5. Calculate the contributions to the chi-square statistic.

6.2.4   Processing the Example

1. Inputting the data:

Input the data manually into a matrix. Assign row names (**rownames**) and column names (**colnames**):

```
> colour<-matrix(c(592,544,119,97,849,677,504,451,36,14),ncol=5)
> rownames(colour)<-c("Boys","Girls")
> colnames(colour)<-c("Fair","Red","Medium","Dark","Jet Black")
```

2. Visualising the data:

The data can be represented as more common in girls than in boys according to sex in one graph window (Figure 6.2). In order to do so, we use the **par** function, (see Section 3.1.6, p. 67) and the attribute which defines the number of graphs by row and column (**mfrow=c(2,1)**):

```
> par(mfrow=c(2,1))
> barplot(colour[1,],main="Boys")
> barplot(colour[2,],main="Girls")
```

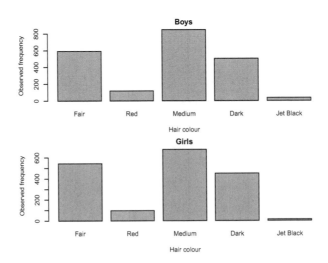

Figure 6.2
Distributions of hair colour by sex.

3. (Optional) Calculating row and column profiles:

To explore the data, it can be interesting to calculate the row and column profiles. First calculate the joint frequencies, that is to say, the percentages n_{ij}/n. The results can be presented by multiplying by 100 and then rounding to the first decimal place using the **round** function:

```
> round(100 * colour / sum(colour), 1)
      Fair Red Medium Dark Jet Black
Boys  15.2 3.1   21.9 13.0        0.9
Girls 14.0 2.5   17.4 11.6        0.4
```

The row profiles $(n_{ij}/\sum_j n_{ij})$ are calculated using the **prop.table** function and the argument **margin = 1**. Note that, to use this function, the dataset must be a matrix. If this is not the case, it must be converted to a matrix using the function **as.matrix**.

```
> round(100 * prop.table (colour, margin = 1), 1)
      Fair Red Medium Dark Jet Black
Boys  28.2 5.7   40.4 24.0        1.7
Girls 30.5 5.4   38.0 25.3        0.8
```

This yields the distributions of hair colour according to sex which are represented in Figure 6.2. There is very little difference between these row profiles. The link between the two variables is not immediately obvious; it can therefore be useful to conduct a test.

The column profiles are calculated in the same way $(n_{ij}/\sum_i n_{ij})$ while specifying that this time we are working on columns (**margin = 2**):

```
> round(100 * prop.table (colour, margin = 2), 1)
      Fair  Red Medium Dark Jet Black
Boys  52.1 55.1   55.6 52.8         72
Girls 47.9 44.9   44.4 47.2         28
```

With regards to the column profiles, the greatest differences are for those with jet black hair: 72% of those with jet black hair are boys.

4. Constructing the chi-square test:

In order to conduct the test of independence between the variables sex and hair colour, we calculate the χ^2_{obs} value and determine the p-value associated with the test:

```
> results <- chisq.test(colour)
> results
```

```
        Pearson's Chi-squared test
```

```
data:   colour
X-squared = 10.4674, df = 4, p-value = 0.03325
```

The p-value indicates that such a big χ^2_{obs} value would have 3.32% probability of being observed in a sample of this size if hair colour were independent of sex. Thus, at the 5% threshold, we reject the independence hypothesis and conclude that, according to this dataset, hair colour depends on sex.

5. Calculating the contributions to the χ^2-statistic:

This relationship between the two variables can be studied in greater detail by calculating the contributions $\frac{(n_{ij}-T_{ij})^2}{T_{ij}}$ to the χ^2_{obs} statistic. The square roots of these contributions are in the object `residuals`. By dividing each term by the total (i.e. the χ^2_{obs} value contained in the `stat` object), we obtain the contributions expressed as percentage:

```
> round(100 * results$residuals^2 / results$stat, 1)
      Fair Red Medium Dark Jet Black
Boys   7.8 0.4    6.5  2.9        28.4
Girls  9.2 0.5    7.7  3.4        33.4
```

The combinations which contribute the most to the non-independence of the two variables are those concerning `Jet Black`. In order to interpret the contribution, we inspect the residuals, and more precisely their positivity or negativity:

```
> round(results$residuals, 3)
          Fair    Red Medium   Dark Jet Black
Boys   -0.903  0.202  0.825 -0.549        1.723
Girls   0.979 -0.219 -0.896  0.596       -1.870
```

It can be said that the number of boys with jet black hair is greater than expected (implied: greater than if the independence hypothesis were true) and there are fewer girls than expected.

6.2.5   Rcmdr Corner

In Rcmdr, it is impossible to construct a chi-square test of independence from a contingency table previously input. If the data is in the form of a contingency table, we recommend that you conduct the test using the command window (see above).

Nevertheless, it is possible:

- To input data directly into Rcmdr and then to analyse that data. In this way, we can obtain the row or column percentages, the contributions and the test statistic:

```
Statistics → Contingency tables
→ Enter and analyze two-way table...
```

- To use the existing data in the form of a table of individuals × variables (see Section 6.2.6, "Taking Things Further"). We then use the menu `Statistics → Contingency tables → Two-way table` to select two variables and obtain a contingency table. This procedure yields the result of the test and the contributions to the χ^2 statistic.

6.2.6 Taking Things Further

The data may arrive from a table of individuals × variables: with the children in rows (i.e. the statistical individual), and the qualitative variables in the columns, which here are `colour` and `sex`. The same procedure can therefore be carried out as before using the function **xtabs** (see Section 2.6, p. 46). Once the contingency table has been constructed, it is possible to analyse it as previously described:

```
> cont.tab <- xtabs(~colour+sex, data=dataset)
> chisq.test(cont.tab)
```

The contingency tables can be visualised with a correspondence analysis (CA) by using the **CA** function (Worked Example 10.2, p. 222) from the FactoMineR package.

Theoretical explanations and examples of the chi-square test are available in many books, such as Clarke and Cooke (2004).

6.3 Comparison of Two Means

6.3.1 Objective

The aim of this worked example is to test the equality of the means of two sub-populations (μ_1 and μ_2) for a quantitative variable X. In formal terms, we test the hypothesis $H_0 : \mu_1 = \mu_2$ against the alternative hypothesis $H_1 : \mu_1 \neq \mu_2$ (or $H_1 : \mu_1 > \mu_2$ or $H_1 : \mu_1 < \mu_2$).

In order to carry out this test, we have to know if the variances of the variable X (σ_1^2 and σ_2^2) in each sub-population are equal or not. Consequently, we must first test the hypothesis $H_0 : \sigma_1^2 = \sigma_2^2$ against the alternative hypothesis $H_1 : \sigma_1^2 \neq \sigma_2^2$. The test statistic is $F = \frac{\hat{\sigma}_1^2}{\hat{\sigma}_2^2}$ and under the null hypothesis this quantity follows a Fisher's distribution with, respectively, $n_1 - 1$ and $n_2 - 1$ degrees of freedom (where n_1 and n_2 are the sample size in each sub-population).

- If we accept the equality of the two variances, we test the equality of the two means using the following t-test. The variance of the difference $(\bar{X}_1 - \bar{X}_2)$ is equal to $\hat{\sigma}_D^2 = \frac{(n_1-1)\hat{\sigma}_1^2 + (n_2-1)\hat{\sigma}_2^2}{n_1+n_2-2} \left(\frac{1}{n_1} + \frac{1}{n_2} \right)$. The test statistic is $T = \frac{\bar{X}_1 - \bar{X}_2}{\hat{\sigma}_D}$ and, under the null hypothesis ($H_0 : \mu_1 = \mu_2$), this quantity follows a Student's t distribution with $n_1 + n_2 - 2$ degrees of freedom.

- If we reject the equality of the two variances, we test the equality of the two means using Welch's t-test. The variance of the difference $(\bar{X}_1 - \bar{X}_2)$ is equal to $\hat{\sigma}_D^2 = \frac{\hat{\sigma}_1^2}{n_1} + \frac{\hat{\sigma}_2^2}{n_2}$. The test statistic is $T = \frac{\bar{X}_1 - \bar{X}_2}{\hat{\sigma}_D}$ and, under the null hypothesis ($H_0 : \mu_1 = \mu_2$), this quantity follows a Student's t distribution with ν degrees of freedom, where $\frac{1}{\nu} = \frac{1}{n_1-1} \left(\frac{\hat{\sigma}_1^2/n_1}{\hat{\sigma}_D^2} \right)^2 + \frac{1}{n_2-1} \left(\frac{\hat{\sigma}_2^2/n_2}{\hat{\sigma}_D^2} \right)^2$.

6.3.2 Example

We want to compare the weights of male and female adult octopuses. We have a dataset with fifteen male and thirteen female octopuses fished off the coast of Mauritania. Table 6.1 features an extract from the dataset.

TABLE 6.1
Extract from the Octopus
Dataset (weight in grams)

Weight	Sex
300	Female
700	Female
850	Female
⋮	⋮
5400	Male

We would like to test the equality of the unknown theoretical mean female (μ_1) and male (μ_2) octopus weights, with a type-one error rate set at 5%.

6.3.3 Steps

1. Read the data.

2. Compare the two sub-populations graphically.

3. Calculate the descriptive statistics (mean, standard deviation, and quartiles) for each sub-population.

4. (Optional) Test the normality of the data in each sub-population.

5. Test the equality of variances.

6. Test the equality of means.

6.3.4 Processing the Example

1. Reading the data:

```
> octopus <- read.table("Octopus.csv",header=T,sep=";")
```

Summary of the dataset:

```
> summary(octopus)
     Weight           Sex
 Min.    : 300    Female:13
 1st Qu.:1480    Male  :15
 Median :1800
 Mean    :2099
 3rd Qu.:2750
 Max.    :5400
```

Using this procedure, we obtain descriptive statistics and we can check that the variable Weight is quantitative and that the variable Sex is qualitative. Be aware that if, for example, the categories of the variable Sex were coded 1 and 2, it would be necessary to transform this variable into a factor prior to any analysis as R would consider it to be quantitative (see Section 1.4.2, p. 7). In order to do so, we would write octopus[,"Sex"] <- factor(octopus[,"Sex"]).

2. Comparing the two sub-populations graphically:

Before any analysis, it may be interesting to visualise the data. Boxplots are used to compare the distribution of weights in each category of the variable Sex:

```
> boxplot(Weight ~ Sex, ylab="Weight", xlab="Sex", data=octopus)
```

Figure 6.3 shows that the males are generally heavier than the females as both weight medians and quartiles are greater for males.

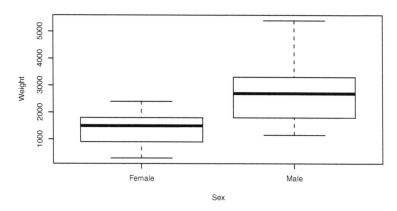

Figure 6.3
Boxplots of octopus weights according to sex.

3. Calculating the descriptive statistics for each sub-population:

Now calculate the mean, standard deviation and quartiles for each sex using the **tapply** function (the argument na.rm = TRUE is in fact useless here as there is no missing data):

```
> tapply(octopus[,"Weight"], octopus[,"Sex"], mean, na.rm=TRUE)
  Female     Male
1405.385 2700.000
> tapply(octopus[,"Weight"], octopus[,"Sex"], sd, na.rm=TRUE)
   Female      Male
 621.9943 1158.3547
> tapply(octopus[,"Weight"],octopus[,"Sex"],quantile,na.rm=TRUE)
$Female
  0%  25%  50%  75% 100%
 300  900 1500 1800 2400

$Male
   0%   25%   50%   75%  100%
 1150 1800 2700 3300 5400
```

4. (Optional) Testing the normality of the data in each sub-population:

In order to construct the comparison of means test, we assume that the mean estimator in each sub-population follows a normal distribution. This is true if the data is normally distributed or if the sample size is large enough (in practice, greater than 30, thanks to the central limit theorem). Here there are less than thirty individuals; therefore data normality must be tested for each sub-population. In order to do so, the Shapiro-Wilk test is used. To test the normality of the males alone, select the weight of the males by requiring that the qualitative variable `Sex` carries the category `Male`. Rows are selected by building the logic vector `select.males`. The components of this vector are `TRUE` for males and are otherwise `FALSE`. We then conduct the Shapiro-Wilk test on the individuals in this selection. Before conducting the Shapiro-Wilk test, we draw the normal QQ-plot that gives an account of probability plotting for informal assessment of normality (see Figure 6.4):

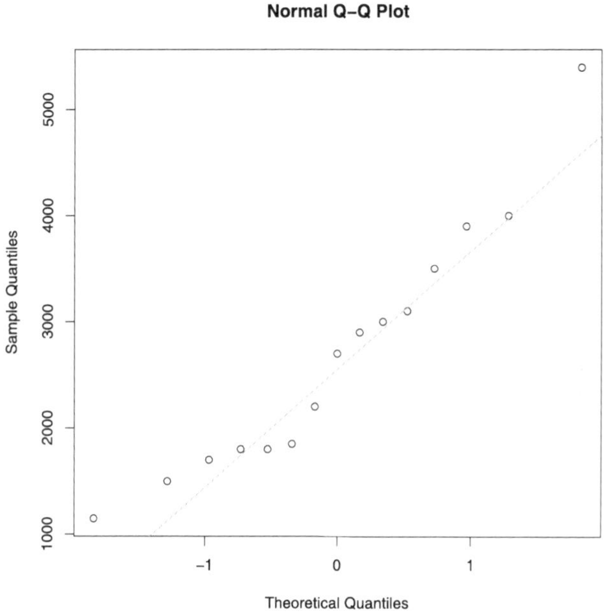

Figure 6.4
Normal QQ-plot for male octopus weights.

```
> select.males <- octopus[,"Sex"]=="Male"
> qqnorm(octopus[select.males,"Weight"])
> qqline(octopus[select.males,"Weight"],col="grey")
> shapiro.test(octopus[select.males,"Weight"])

        Shapiro-Wilk normality test

data:  octopus[select.males, "Weight"]
W = 0.935, p-value = 0.3238
```

As the p-value associated with the test is greater than 5%, the normality of the males' weights is accepted. We will not provide the output for the females, but the assumption of normality is also accepted.

When the assumption of normality is rejected, the test of equality of means can be conducted using non-parametric tests such as that of Wilcoxon (**wilcox.test**) or Kruskal-Wallis (**kruskal.test**).

5. Testing the equality of variances:

In order to compare the mean of the two male and female sub-populations, there are two possible types of tests: one in which the unknown variances of the two sub-populations are different and the other in which they are equal. We must therefore test the equality of variances $H_0 : \sigma_1^2 = \sigma_2^2$ against $H_1 : \sigma_1^2 \neq \sigma_2^2$ using an F-test:

```
> var.test(Weight ~ Sex, conf.level=.95,data=octopus)

        F test to compare two variances

data:  Weight by Sex
F = 0.2883, num df = 12, denom df = 14, p-value = 0.03713
alternative hypothesis: true ratio of variances is
                        not equal to 1
95 percent confidence interval:
 0.0945296 0.9244467
sample estimates:
ratio of variances
          0.2883299
```

The p-value associated with the test of comparison of variances is 0.037: H_0 can therefore be rejected and we can thus consider that the variances are significantly different. The variance ratio ($\hat{\sigma}_1/\hat{\sigma}_2$) is 0.288 and the confidence interval for this ratio (to 95%) is [0.09 ; 0.92]. In the variance ratio, the numerator is the variance of the weights at the first level of the variable Sex alphabetically (here, Female) and the denominator is the variance of the weights at the second level of the variable Sex.

6. Testing the equality of means:

We here use the **t.test** function. As the variances are different, Welch's test is used to compare the means. In order to do so, specify that the variances are unequal using the argument var.equal=FALSE. If the variance were equal, we would have used a Student's t-test with the argument var.equal=TRUE. The default test is bilateral (alternative='two.sided'), however the alternative hypothesis H_1 may be that the males are lighter (alternative="less") or heavier (alternative="greater"). We here consider the category of reference to be Female (first level of the variable Sex) and the other category is tested according to this reference.

```
> t.test(Weight~Sex, alternative='two.sided', conf.level=.95,
      var.equal=FALSE, data=octopus)

         Welch Two Sample t-test

data:  Weight by Sex
t = -3.7496, df = 22.021, p-value = 0.001107
alternative hypothesis: true difference in means
                        is not equal to 0
95 percent confidence interval:
 -2010.624  -578.607
sample estimates:
mean in group Female   mean in group Male
           1405.385            2700.000
```

The p-value (0.001) associated with the test of unequal variances indicates that the means are significantly different. The mean weight for males within the population (estimated at 2700 g) is thus significantly different from that of females (estimated at 1405 g).

6.3.5 Rcmdr Corner

1. Reading the data from a file:

Data → Import data → from text file, clipboard, or URL ...

It must then be specified that the column separator is ";".
To check that the dataset has been imported successfully:

Statistics → Summaries → Active data set

2. Comparing the two sub-populations graphically:

Graphs → Boxplot...

Then click on Plot by groups... to obtain one box for each category of the qualitative variable (Identify outliers with mouse is used to identify the extreme individuals, i.e. those beyond the boxplots).

3. Calculating the descriptive statistics for each sub-population:

Statistics → Summaries → Numerical summaries...

Then click on Summarize by groups...

4. (Optional) Testing the normality of the data in each sub-population:

Be aware that the test for normality is conducted for the overall data and cannot be conducted directly on a subset. It is therefore necessary to define the subsets.

Data → Active data set → Subset active data set...

The subset is determined using `Subset expression` (type `Sex == "Male"`) and it is essential to rename the dataset in order not to lose the original (`Name for new data set`).

We can then conduct the Shapiro-Wilk test:

`Statistics → Summaries → Shapiro-Wilk test of normality...`

We then come back to the original data in order to conduct the same analysis for the females: We again select the initial dataset as active by clicking on `Data`. We conduct the test of normality for the females (in the same way as for the males). We then select the initial dataset.

In summary, all of this is somewhat restrictive and it is preferable to conduct these tests via a command line.

5. Testing the equality of variances:

`Statistics → Variances → Two-variances F-test...`

Choose the qualitative variable (`Groups (pick one)`), the quantitative variable (`Response variable (pick one)`) and the alternative test hypothesis (by default, the alternative hypothesis is that the variances are different and the test is `Two-sided`).

6. Testing the equality of means:

`Statistics → Means → Independent samples t-test...`

Choose the qualitative variable (`Groups (pick one)`), the quantitative variable (`Response variable (pick one)`) and the alternative test hypothesis (by default, the alternative hypothesis is that the means are different and the test is `Two-sided`). Specify whether you consider the variances to be equal or unequal (`Assume equal variances?`).

6.3.6   Taking Things Further

If we need to compare more than two means, a number of tests are available. The choice of test depends whether or not the variances in each sub-population are equal. If the variances are equal, refer to the worked example on one-way analysis of variance (p. 157); if the variances are unequal, use the function **oneway.test** (Welch, 1951). Bartlett's test (**bartlett.test**) can be used to test equality of variances.

Both the tests of comparisons of variance and of means are explained in detail in most books such as Clarke and Cooke (2004).

6.4 Testing Conformity of a Proportion

6.4.1 Objective

The aim of this worked example is to test the equality of a proportion π at a given value π_0. If P is the estimator of π and n the sample size, nP follows a binomial distribution with parameters n and π. This distribution allows one to perform an exact test about the probability of success in a Bernoulli experiment.

This test can also be used to test the equality of two proportions. Indeed, testing $\pi_1 = \pi_2$ remains to test $\pi_1 = 0.5$.

6.4.2 Example

We are interested in the intention to vote for a candidate A in the second round of presidential elections. In a poll of 1040 voters, candidate A wins 52.4% of the vote. Can we consider to the 95% threshold that this candidate will win the election? Implicitly, we assume that the sample is representative of the population as a whole, drawn with replacement, and that the voters will vote as predicted by the poll. However, simple random polls use samples drawn without replacement. Nevertheless, the procedure explained below is a reasonable approximation when the poll rate is low, that is, when the size of the population is much greater than that of the sample.

6.4.3 Step

Test the equality of the proportion to 50% with a type-one error rate of 5%.

6.4.4 Processing the Example

As an alternative hypothesis to the test of the equality of a proportion of 50%, we choose the percentage of voting intentions to be greater than 50% (alternative="greater"). The success rate (i.e. the number of votes for candidate A) must be a whole number. We therefore round the decimal yielded by the problem to the nearest whole number (using the **round** function with 0 decimal digits). We then use the **binom.test** function, but the **prop.test** function yields exactly the same result:

```
> nbr.vot.A <- round(0.524 * 1040,0)
> binom.test(nbr.vot.A, n=1040, p=0.5, alternative="greater")
        Exact binomial test

data:   nbr.vot.A and 1040
nb of successes = 545, nb of trials = 1040, p-value = 0.06431
```

```
alternative hypothesis: true probability of success is > 0.5
95 percent confidence interval:
 0.4980579 1.0000000
sample estimates:
probability of success
              0.5240385
```

As the *p*-value is greater than 5%, we accept the hypothesis that the proportion is equal to 50%: we cannot therefore confirm from this poll that candidate A will win the election.

The previous exact binomial test yields a confidence interval for the proportion at the 95% level. However, since we have constructed a unilateral test, this interval places all the error (i.e. 5%) on one side. More precisely, the interval provided is of the format $[a, 1]$ where a is determined so as $\Pr(a \leq p \leq 1) \geq 0.95$. To obtain a confidence interval of the format $[a, b]$, with approximately 2.5% on each side, a bilateral test is required (two.sided), and this is the default option:

> **binom.test**(nbr.vot.A,n=1040)

```
            Exact binomial test

data:  nbr.vot.A and 1040
number of successes=545, number of trials=1040, p-value=0.1286
alternative hypothesis: true probability of success
                       is not equal to 0.5
95 percent confidence interval:
 0.4931733 0.5547673
sample estimates:
probability of success
              0.5240385
```

Strictly speaking, the unknown probability p estimated at 0.524 either does or does not belong to the interval $[0.493, 0.555]$. It is therefore within the interval with a probability of 0 or 1. However, the overall procedure guarantees that, if it were to be repeated infinitely with new samples of the same size $n = 1040$ participants, 95% of the confidence intervals would contain the true unknown value p. In practice, even if that is not the case, it is quicker to say that the percentage of voting intentions is between 49.3% and 55.5%.

6.4.5   Rcmdr Corner

There is no test for the conformity of a proportion in Rcmdr.

6.4.6   Taking Things Further

The tests of conformity to a proportion are explained in detail in most books such as Moore et al. (2007).

6.5   Comparing Several Proportions

6.5.1   Objective

The aim of this worked example is to test the equality of several proportions.

6.5.2   Example

We shall again examine the example of hair colour for boys and girls in a Scottish county (see Worked Example 6.2, p. 113):

	Fair	Red	Medium	Dark	Jet Black
Boys	592	119	849	504	36
Girls	544	97	677	451	14

We would like to compare the proportions of boys for different groups, which here is different hair colours. We will test whether or not these proportions are equal in all the groups, with a type-one error set at 5%.

6.5.3   Step

Test the equality of proportions for boys (for different hair colours).

6.5.4   Processing the Example

To conduct the test of equality of proportions, we provide a list with the number of boys for each hair colour, then a list with the total number of individuals for each hair colour:

```
> prop.test(c(592,119,849,504,36),n=c(1136,216,1526,955,50))

    5-sample test for equality of proportions without continuity
    correction

data: c(592,119,849,504,36) out of c(1136,216,1526,955,50)
X-squared = 10.4674, df = 4, p-value = 0.03325
alternative hypothesis: two.sided
sample estimates:
   prop 1    prop 2    prop 3    prop 4    prop 5
0.5211268 0.5509259 0.5563565 0.5277487 0.7200000
```

Significant differences between proportions can thus be confirmed (the *p*-value is lower than 5%). We can see that the proportion of boys estimated for the jet black group is much greater than for the other groups. Note that the test which is performed is a chi-square test of independence and the results are exactly the same as the ones obtained in the worked example on chi-square (p. 113).

6.5.5   Rcmdr Corner

There is no test of comparison of proportions in Rcmdr.

6.5.6   Taking Things Further

The **prop.test** function can also be applied in cases where there are two proportions (corresponding to two variables with two categories) as Yates' correction for continuity is offered by default in this situation.

Tests for equality of proportions are explained in detail in books such as Moore et al. (2007).

6.6   The Power of a Test

6.6.1   Objective

The aim of this worked example is to calculate the power of the test of equality of means of two sub-populations. The test of equality of means is used to choose between the hypothesis $H_0 : \mu_1 = \mu_2$ and the alternative hypothesis $H_1 : \mu_1 \neq \mu_2$ (or $H_1 : \mu_1 > \mu_2$ or $H_1 : \mu_1 < \mu_2$) (see Worked Example 6.3). The power of the test is the probability of rejecting the hypothesis H_0 when H_1 is true. Power is equal to $1 - \beta$, β being the type-two error (that is to say, the risk of mistakenly accepting H_0). The power therefore corresponds to the probability of detecting a difference in means, if indeed this difference exists. The advantage of calculating the power of a test prior to conducting an experiment lies mainly in the ability to optimise the number of trials (i.e. statistical individuals) according to the aim of the experimenter. Indeed, the power of the test is directly related to the number of individuals per group (n), the amplitude of the difference that we want to detect (δ), the within-group variability (σ), and the type-one error (α).

6.6.2   Example

We now examine the example of an experiment in milk production. Researchers at INRA (the French National Institute for Agricultural Research) selected two genetically different types of dairy cow according to the volume of milk produced. The aim is to detect a potential difference in the protein levels in the milks produced by these two sub-populations.

During a previous study, the standard deviation of protein levels in the milk from a herd of Normandy cows was found to be 1.7 g/kg of milk. As an approximation we will therefore use the standard deviation $\sigma = 1.7$ and use the classical $\alpha = 5\%$ threshold. The aim is to have $\beta = 80\%$ chance of detecting a difference in the means of the protein levels of $\delta = 1$ g/kg of milk from the two populations. To meet this objective, we will determine the number of dairy cows required using the function **power.t.test**.

6.6.3   Steps

1. Calculate the number of individuals required to obtain a power of 80%.

2. Calculate the power of the test with twenty individuals per group.

3. Calculate the difference of means detectable at 80% with twenty individuals per group.

6.6.4   Processing the Example

1. Calculating the number of individuals required to obtain a power of 80%:

```
> power.t.test(delta=1, sd=1.7, sig.level=0.05, power=0.8)
     Two-sample t test power calculation

              n = 46.34674
          delta = 1
             sd = 1.7
      sig.level = 0.05
          power = 0.8
    alternative = two.sided

  NOTE: n is number in *each* group
```

Here we can see that a minimum of 47 individuals are required per population in order to have more than an 80% chance of detecting a difference in means in the protein levels of 1 g/kg of milk between the two populations.

2. Calculating the power of the test with twenty individuals per group:

```
> power.t.test(n=20, delta=1, sd=1.7, sig.level=0.05)$power
[1] 0.4416243
```

If we decide to use only twenty cows per population in the experiment, there will be a 44% chance of detecting a difference in means in the protein levels of 1 g/kg of milk between the two populations.

3. Calculating the difference detectable at 80% with twenty individuals per group:

```
> power.t.test(n=20, sd=1.7, sig.level=0.05, power=0.8)$delta
[1] 1.545522
```

If we decide to use only twenty cows per population in the experiment, there will then be an 80% chance of detecting a difference in means in the protein levels of 1.55 g/kg of milk between the two populations.

6.6.5   Rcmdr Corner

It is not possible to calculate the power of a test with Rcmdr.

6.6.6   Taking Things Further

If we want to compare more than two means, and if we suppose that the variances in each sub-population are equal, we can construct a one-way analysis of variance model (see p. 157). In order to calculate the power of the test beforehand, use function **power.anova.test** and then test the overall effect. In order to do so, we must first have an idea of the within-group variance (or residual mean square) which will be observed, as well as defining the between-group variance (or mean square associated to the factor) that we wish to detect.

   We examine the same example, but this time consider that there are three genetic populations. We again assume that the residual standard deviation of the protein level is 1.7 g/kg of milk and that we want to detect (with a threshold 5%) whether there really is a difference between the three populations if the true means of protein levels are, for example, 28, 30 and 31 g/kg, respectively.

```
> 1.7^2
[1] 2.89
> var(c(28, 30, 31))
[1] 2.333333
> power.anova.test(groups=3, between.var=2.3333,
    within.var=2.89, power=.80)

    Balanced one-way analysis of variance power calculation

        groups = 3
             n = 7.067653
    between.var = 2.3333
     within.var = 2.89
     sig.level = 0.05
         power = 0.8

 NOTE: n is number in each group
```

The foreseen residual variance is 2.89 g^2/kg^2 and the between-group variance that we want to detect as significant is 2.33 g^2/kg^2. We consider that eight cows are therefore needed per group in order to have an 80% chance of detecting the effect of the genetical population on the protein level of the milk.

The power of a t-test for mean comparison is explained in books such as Clarke and Cooke (2004). The power of an F-test in analysis of variance is described in Sahai and Ageel (2000), for example.

7

Regression

7.1 Simple Linear Regression

7.1.1 Objective

Simple linear regression is a statistical method used to model the linear relationship between two quantitative variables for explanatory or prediction purposes. Here we have one explanatory variable (denoted X) and one response variable (denoted Y), connected by the following model:

$$Y = \beta_0 + \beta_1 X + \varepsilon$$

where ε is the variable for noise or measurement error rate. The parameters β_0 and β_1 are unknown. The aim is to estimate them from a sample of n pairs $(x_1, y_1), \ldots, (x_n, y_n)$. The model is written in indexed form:

$$y_i = \beta_0 + \beta_1 x_i + \varepsilon_i$$

The β_0 coefficient corresponds to the intercept and β_1 to the slope. We estimate these parameters by minimising the least-square criterion

$$(\hat{\beta}_0, \hat{\beta}_1) = \operatorname*{argmin}_{\beta_0, \beta_1} \sum_{i=1}^{n} (y_i - \beta_0 - \beta_1 x_i)^2$$

Once the parameters have been estimated ($\hat{\beta}_0$ and $\hat{\beta}_1$), we obtain the regression line:

$$f(x) = \hat{\beta}_0 + \hat{\beta}_1 x$$

from which predictions can be made. The adjusted or smoothed values are defined by

$$\hat{y}_i = \hat{\beta}_0 + \hat{\beta}_1 x_i$$

and the residuals by

$$\hat{\varepsilon}_i = y_i - \hat{y}_i$$

Analysing the residuals is essential as it is used to check the individual fitting (outlier) and the global fitting of the model, for example by checking that there is no structure.

7.1.2 Example

Air pollution is currently one of the most serious public health worries world-wide. Many epidemiological studies have proved the influence that some chemical compounds such as sulphur dioxide (SO_2), nitrogen dioxide (NO_2), ozone (O_3) or other air-borne dust particles can have on our health.

Associations set up to monitor air quality are active all over France to measure the concentration of these pollutants. They also keep a record of meteorological conditions such as temperature, cloud cover, wind, etc.

Here we analyse the relationship between the maximum daily ozone level (in $\mu g/m^3$) and temperature. We have at our disposal 112 observations collected during the summer of 2001 in Rennes (France).

7.1.3 Steps

1. Read the data.

2. Represent the scatterplot (x_i, y_i).

3. Estimate the parameters.

4. Draw the regression line.

5. Conduct a residual analysis.

6. Predict a new value.

7.1.4 Processing the Example

1. Reading the data:
Read the data and summarise the variable of interest, here max03 and T12:

```
> ozone <- read.table("ozone.txt",header=T)
> summary(ozone[,c("maxO3","T12")])
     maxO3              T12
 Min.   : 42.00    Min.   :14.00
 1st Qu.: 70.75    1st Qu.:18.60
 Median : 81.50    Median :20.55
 Mean   : 90.30    Mean   :21.53
 3rd Qu.:106.00    3rd Qu.:23.55
 Max.   :166.00    Max.   :33.50
```

2. Representing the scatterplot (x_i, y_i):

```
> plot(maxO3~T12,data=ozone,pch=15,cex=.5)
```

Each point on the graph (Figure 7.1) represents, for a given day, a temperature measurement taken at midday, and the ozone peak for the day. From this graph, the relationship between temperature and ozone concentration seems to be linear.

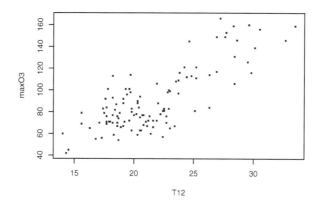

Figure 7.1
Representation of pairs (x_i, y_i) with a scatterplot.

3. Estimating the parameters:
The **lm** function (linear model) is used to fit a linear model:

```
> simple.reg <- lm(max03~T12,data=ozone)
> summary(simple.reg)
Call:
lm(formula = max03 ~ T12, data = ozone)

Residuals:
     Min        1Q    Median        3Q       Max
-38.0789  -12.7352   0.2567   11.0029   44.6714

Coefficients:
            Estimate Std. Error  t value   Pr(>|t|)
(Intercept) -27.4196     9.0335   -3.035    0.003 **
T12           5.4687     0.4125   13.258   <2e-16 ***
---
Signif. codes:  0 '***' 0.001 '**' 0.01 '*' 0.05 '.' 0.1 ' ' 1

Residual standard error: 17.57  on 110  degrees of freedom
Multiple R-Squared: 0.6151,     Adjusted R-squared: 0.6116
F-statistic: 175.8 on 1 and 110 DF, p-value: < 2.2e-16
```

Amongst other things, we obtain a `Coefficient` matrix which, for each parameter (each row), has four columns: its estimation (`Estimate` column), its estimated standard deviation (`Std. Error`), the observed value statistic for the test $H_0 : \beta_j = 0$ against $H_1 : \beta_j \neq 0$. Finally, the p-value (`Pr(>|t|)`) yields the probability to obtain such a large statistic under H_0.

Coefficients β_0 and β_1 are estimated by -27.4 and 5.5. The test of significance for the coefficients here gives p-values of 0.003 and around $2.\,10^{-16}$. Thus, the null hypothesis for each of these tests is rejected in favour of the alternative hypothesis. The p-value of less than 5% for the constant (intercept) indicates that the constant must appear in the model. The p-value less than 5% for the slope indicates a significant link between `max03` and `T12`.

The summary of the estimation step features the estimation of residual standard deviation σ, which here is 17.57, as well as the corresponding number of degrees of freedom $n - 2 = 110$.

The value of R^2 and adjusted R^2 are also given. The value of R^2 is rather high ($R^2 = 0.6151$), thus supporting the suspected linear relationship between the two variables. In other words, 61% of the variability of `max03` is explained by `T12`.

The final row, which is particularly useful for multiple regression, indicates the result of the test of the comparison between the model used and the model using only the constant as explanatory variable.

We can consult the list of different results (components of the list, see Section 1.4.7, p. 20) of the object `simple.reg` using

```
> names(simple.reg)
 [1] "coefficients"  "residuals"   "effects"    "rank"
 [5] "fitted.values" "assign"      "qr"         "df.residual"
 [9] "xlevels"       "call"        "terms"      "model"
```

We can then retrieve the coefficients using

```
> simple.reg$coef
(Intercept)          T12
 -27.419636    5.468685
```

or by using the **coef** function:

```
> coef(simple.reg)
```

To fit the model without the constant, we proceed as follows:

```
> reg.without.intercept <- lm(max03~T12-1, data=ozone)
```

4. Drawing the regression line:

We can simply apply the following command to obtain Figure 7.2:

```
> plot(maxO3~T12,data=ozone,pch=15,cex=.5)
> abline(simple.reg)
```

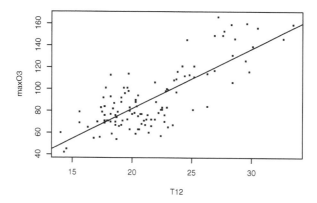

Figure 7.2
Scatterplot and regression line.

5. Conducting a residual analysis:
The residuals are obtained using the **residuals** function; however, these residuals do not have the same variance (heteroscedastic). We therefore need to use and plot (Figure 7.3) studentised residuals, which have the same variance.

```
> res.simple<-rstudent(simple.reg)
> plot(res.simple,pch=15,cex=.5,ylab="Residuals")
> abline(h=c(-2,0,2),lty=c(2,1,2))
```

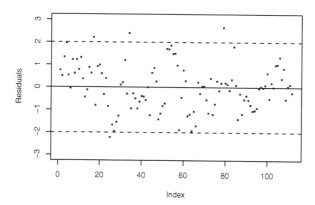

Figure 7.3
Representation of residuals.

In theory, 95% of the studentised residuals can be found in the interval $[-2, 2]$. This is the case here because only four residuals are found outside this interval.

6. Predicting a new value:

As we now have a new observation xnew, we can simply use the estimations to predict the corresponding Y value. However, the predicted value is of little interest without its corresponding confidence interval. Let us examine an example of this. Here, we have a new temperature observation T12 equal to 19 degrees for 1st October 2001.

```
> xnew <- 19
> xnew <- as.data.frame(xnew)
> colnames(xnew) <- "T12"
> predict(simple.reg,xnew,interval="pred")
         fit      lwr      upr
[1,] 76.48538 41.4547 111.5161
```

It must be noted that the xnew argument of the **predict** function must be a data-frame in which the names of the explanatory variables are the same as in the original dataset (here T12). The predicted value is 76.5 and the prediction interval at 95% is $[41.5, 111.5]$. To represent, on one single graph, the confidence interval of a fitted value along with that of a prediction, we calculate these intervals for all the points which were used to draw the regression line. We make the two appear on the same graph (Figure 7.4).

```
> gridx <- data.frame(T12=seq(min(ozone[,"T12"]),
      max(ozone[,"T12"]),length=100))
> CIline <- predict(simple.reg,new=gridx,
      interval="conf",level=0.95)
> CIpred <- predict(simple.reg,new=gridx,
      interval="pred",level=0.95)
> plot(maxO3~T12,data=ozone,pch=15,cex=.5)
> matlines(gridx,cbind(CIline,CIpred[,-1]),
      lty=c(1,2,2,3,3),col=1)
> legend("topleft",lty=2:3,c("pred","conf"))
```

7.1.5 Rcmdr Corner

1. Reading the data from a file:

Data → Import data → from text file, clipboard, or URL ...

Next specify that the field separator is the space and the decimal-point character is ".".

To check that the dataset has been imported successfully:

Statistics → Summaries → Active data set

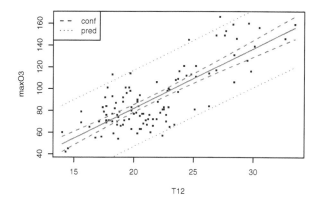

Figure 7.4
Confidence and prediction intervals.

2. Representing the scatterplot graphically (x_i, y_i):
Graphs → Scatterplot... Then select variables x and y. It is possible to identify specific points using the mouse. By default, a boxplot is drawn for each variable x and y, the regression line is drawn as well as a smoothing curve, but it is possible to uncheck these options in the dialogue box.

3. Estimating the parameters:
Statistics → Fit models → Linear regression... then select the explanatory variable (x) and the response variable (y). It is then possible to obtain a confidence interval for each parameter of the regression (β_0 and β_1):

Models → Confidence intervals...

4. Drawing the regression line:
The regression line is drawn when we create the cloud of points (see item 2).

5. Conducting a residual analysis:
Models → Add observation statistics to data...

We can then add to the dataset the fitted values, the residuals, the studentised residuals, the estimated values and Cook's distances. We can then construct graphs using these new variables. Many graphs are available directly:

Models → Graphs → Basic diagnostic plots

6. Predicting a new value:
It is not possible to predict a new value automatically with Rcmdr.

7.1.6 Taking Things Further

Theoretical reminders about linear regression are available in many books, such as Clarke and Cooke (2004), Moore et al. (2007) or Faraway (2005).

7.2 Multiple Linear Regression

7.2.1 Objective

Multiple linear regression consists of explaining and/or predicting a quantitative variable Y by p quantitative variables $X_1, \cdots, X_p$. The multiple regression model is a generalisation of the simple regression model (see Worked Example 7.1). We therefore suppose that the n data collected conform to the following model:

$$y_i = \beta_0 + \beta_1 x_{i1} + \beta_2 x_{i2} + \cdots + \beta_p x_{ip} + \varepsilon_i, \qquad i = 1, \cdots, n, \qquad (7.1)$$

where x_{ij} are known numbers measuring the explanatory variables which are organised within a matrix X known as the experimental design matrix. The parameters β_j of the model are unknown and need to be estimated. The parameter β_0 (intercept) corresponds to the constant of the model. The ε_i are random unknown variables and represent errors of measurement.

By writing the model 7.1 in the form of a matrix, we obtain

$$\mathbb{Y} = \mathbb{X}\,\beta + \varepsilon \qquad (7.2)$$

$$\mathbb{Y} = \begin{pmatrix} y_1 \\ \vdots \\ y_n \end{pmatrix}, \quad \mathbb{X} = \begin{pmatrix} 1 & x_{11} & \cdots & x_{1p} \\ \vdots & \vdots & \vdots & \vdots \\ 1 & x_{n1} & \cdots & x_{np} \end{pmatrix}, \quad \beta = \begin{pmatrix} \beta_0 \\ \vdots \\ \beta_p \end{pmatrix}, \quad \varepsilon = \begin{pmatrix} \varepsilon_1 \\ \vdots \\ \varepsilon_n \end{pmatrix}$$

From these observations, we estimate the unknown parameters of the model by minimising the least-square criterion:

$$\hat{\beta} = \operatorname*{argmin}_{\beta_0, \cdots, \beta_p} \sum_{i=1}^{n} \left(y_i - \beta_0 - \sum_{j=1}^{p} \beta_j x_{ij} \right)^2 = \operatorname*{argmin}_{\beta \in \mathbb{R}^{p+1}} (\mathbb{Y} - \mathbb{X}\beta)'(\mathbb{Y} - \mathbb{X}\beta)$$

If matrix X is full rank, that is to say, if the explanatory variables are not collinear, the least square estimator $\hat{\beta}$ of β is

$$\hat{\beta} = (\mathbb{X}'\mathbb{X})^{-1}\mathbb{X}'\mathbb{Y}$$

Once the parameters have been estimated, we can calculate the fitted values:

$$\hat{y}_i = \hat{\beta}_0 + \hat{\beta}_1 x_{i1} + \cdots + \hat{\beta}_p x_{ip}$$

or predict new values. The difference between the observed value and the fitted value is, by definition, the residual:

$$\hat{\varepsilon}_i = y_i - \hat{y}_i$$

Analysing the residuals is essential as it is used to check the individual fitting of the model (outlier) and the global fitting, for example by checking that there is no structure.

7.2.2 Example

We reexamine the ozone dataset introduced in Worked Example 7.1 (p. 133). Here we analyse the relationship between the maximum daily ozone level (in $\mu g/m^3$) and temperature at different times of day, cloud cover at different times of day, the wind projection on the East-West axis at different times of day and the maximum ozone concentration for the day before the day in question. The data was collected during the summer of 2001 and the sample size is 112.

7.2.3 Steps

1. Read the data.

2. Represent the variables.

3. Estimate the parameters.

4. Choose the variables.

5. Conduct a residual analysis.

6. Predict a new value.

7.2.4 Processing the Example

1. Reading the data:
Read the data and choose the variables of interest, here the response variable, maximum ozone, denoted **maxO3**, and the explanatory variables: temperature, cloud cover, wind projection on the East-West axis at 9am, midday, and 3pm, as well as the ozone maximum from the previous day **maxO3y**.

```
> ozone <- read.table("ozone.txt",header=T)
> ozone.m <- ozone[,1:11]
> names(ozone.m)
 [1] "maxO3"   "T9"       "T12"      "T15"      "Ne9"      "Ne12"
 [7] "Ne15"    "Wx9"      "Wx12"     "Wx15"     "maxO3y"
```

Names are attributed using the **names** or **colnames** function. It can be interesting to summarise these variables (**summary**) or to represent them.

2. Representing the variables:
In order to make sure that there are no errors during input, it is important to conduct a univariate analysis of each variable (histogram, for example). When there are not many variables, they can be represented two by two on the same graph using the **pairs** function (here **pairs(ozone.m)**). It is also possible to explore the data using a principal component analysis with the illustrative explanatory variable (see Worked Example 10.1, p. 209).

3. Estimating the parameters:

To estimate the parameters, we must first write the model. Here, we use **lm** (linear model) with a formula (see Appendix A.2, p. 266), as with simple regression. Here, there are ten explanatory variables and it would therefore be fastidious to write all of them. R software can be used to write them quickly, using the command

```
> reg.mul <- lm(max03~.,data=ozone.m)
> summary(reg.mul)

Call:
lm(formula = max03 ~ ., data = ozone.m)

Residuals:
     Min      1Q  Median      3Q     Max
 -53.566  -8.727  -0.403   7.599  39.458

Coefficients:
              Estimate Std. Error t value Pr(>|t|)
(Intercept)  12.24442   13.47190   0.909   0.3656
T9           -0.01901    1.12515  -0.017   0.9866
T12           2.22115    1.43294   1.550   0.1243
T15           0.55853    1.14464   0.488   0.6266
Ne9          -2.18909    0.93824  -2.333   0.0216 *
Ne12         -0.42102    1.36766  -0.308   0.7588
Ne15          0.18373    1.00279   0.183   0.8550
Wx9           0.94791    0.91228   1.039   0.3013
Wx12          0.03120    1.05523   0.030   0.9765
Wx15          0.41859    0.91568   0.457   0.6486
max03y        0.35198    0.06289   5.597 1.88e-07 ***
---
Signif. codes:  0 '***' 0.001 '**' 0.01 '*' 0.05 '.' 0.1 ' ' 1

Residual standard error: 14.36 on 101 degrees of freedom
Multiple R-Squared: 0.7638,     Adjusted R-squared: 0.7405
F-statistic: 32.67 on 10 and 101 DF, p-value: < 2.2e-16
```

As in simple linear regression (see Worked Example 7.1), outputs from the software are in the form of a matrix (under the name `Coefficients`) with four columns for each parameter: its estimate (column `Estimate`), its estimated standard deviation (`Std. Error`), the observed value of the test statistic for $H_0 : \beta_i = 0$ against $H_1 : \beta_i \neq 0$, and the p-value (`Pr(>|t|)`). The latter gives, as a test statistic under H_0, the probability of exceeding the estimated value.

Here, when they are tested one at a time, the significant variables are `max03y` and `Ne9`. However, since the regression is multiple and the explanatory variables are not orthogonal, it can be misleading to use these tests. Indeed, testing a coefficient means testing the significance of a variable whereas the other variables are in the model. In other words, this means testing that the variable does not provide any additional information, bearing in mind that all the other variables are within the model. It is therefore important to use procedures for choosing the models such as those presented below.

The summary of the estimation step features the estimation of residual standard deviation σ, which here is 14.36, as well as the number of associated degrees of freedom $n - 11 = 101$.

The last row indicates the result of the test of comparison between the model used and the model which uses only the constant as the explanatory variable. It is, of course, significant, as the explanatory variables provide information about the response variable.

4. Choosing the variables:
It is possible to choose the variables by hand, step by step. We remove the least significant variable, T9, and then recalculate the estimations, and so on. In R the leaps package processes the choice of variables: The **regsubsets** function yields, for different criteria (Bayesian information criterion or BIC, adjusted R^2, Mallows' Cp, etc.), the best model (if `nbest=1`) with one explanatory variable, two explanatory variables, ... with `nvmax` explanatory variables. The graphical representation helps to analyse the results.

```
> library(leaps)
> choice <- regsubsets(maxO3~.,data=ozone.m,nbest=1,nvmax=11)
> plot(choice,scale="bic")
```

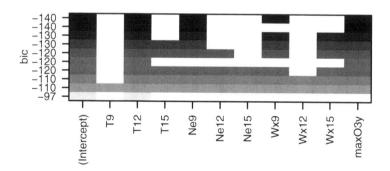

Figure 7.5
Choosing variables with BIC.

We can also determine the variables for the best model according to the BIC criterion using the following line of code:

```
> summary(choice)$which[which.min(summary(choice)$bic),]
(Intercept)        T9       T12       T15       Ne9      Ne12
       TRUE     FALSE      TRUE     FALSE      TRUE     FALSE
       Ne15       Wx9      Wx12      Wx15    max03y
      FALSE      TRUE     FALSE     FALSE      TRUE
```

The criterion is optimal for the top row of the graph. For the Bayesian information criterion (argument `scale="bic"`), we retain the model with four variables: T12, Ne9, Wx9 and max03y. We thus fit the model using the selected variables:

```
> final.reg <- lm(max03~T12+Ne9+Wx9+max03y,data=ozone.m)
> summary(final.reg)

Call:
lm(formula = max03 ~ T12 + Ne9 + Wx9 + max03y, data = ozone.m)

Residuals:
    Min      1Q  Median      3Q     Max
-52.396  -8.377  -1.086   7.951  40.933

Coefficients:
            Estimate Std. Error t value Pr(>|t|)
(Intercept) 12.63131   11.00088   1.148 0.253443
T12          2.76409    0.47450   5.825 6.07e-08 ***
Ne9         -2.51540    0.67585  -3.722 0.000317 ***
Wx9          1.29286    0.60218   2.147 0.034055 *
max03y       0.35483    0.05789   6.130 1.50e-08 ***
---
Signif. codes:  0 '***' 0.001 '**' 0.01 '*' 0.05 '.' 0.1 ' ' 1

Residual standard error: 14 on 107 degrees of freedom
Multiple R-Squared: 0.7622,     Adjusted R-squared: 0.7533
F-statistic: 85.75 on 4 and 107 DF,  p-value: < 2.2e-16
```

5. Conducting a residual analysis:
The observed residuals can be calculated using the **residuals** function on the output of the **lm** function (**residuals(final.reg)**). However, these residuals do not have the same variance (heteroscedastic). We therefore need to use studentised residuals, which have the same variance:

```
> res.m <- rstudent(final.reg)
> plot(res.m,pch=15,cex=.5,ylab="Residuals")
> abline(h=c(-2,0,2),lty=c(2,1,2))
```

In theory, 95% of the studentised residuals can be found in the interval $[-2, 2]$.

This is the case here where only three residuals are found outside this interval (see Figure 7.6).

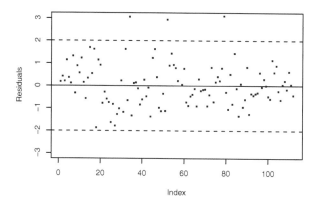

Figure 7.6
Representation of residuals.

6. Predicting a new value:

As we now have a new observation, we can simply use the estimators to predict the corresponding Y value. However, the predicted value is of little interest without its corresponding confidence interval. Let us examine an example of this. For 1st October 2001, we obtain the following values: max03y=70, T12=19, Ne9=8 and Wx9=2.05. It must be noted that the xnew argument of the **predict** function must be a data-frame with the same names as the explanatory variables of the original dataset.

```
> xnew <- matrix(c(19,8,2.05,70),nrow=1)
> colnames(xnew) <- c("T12","Ne9","Wx9","max03y")
> xnew <- as.data.frame(xnew)
> predict(final.reg,xnew,interval="pred")
          fit      lwr      upr
[1,] 72.51437 43.80638 101.2224
```

The predicted value is 72.5 and the confidence interval at 95% for the prediction is $[43.8, 101.2]$.

7.2.5 Rcmdr Corner

1. Reading the data from a file:

Data → Import data → from text file, clipboard, or URL ...

Next, specify that the field separator is the space and the decimal-point character is ".".

To check that the dataset has been imported successfully:
`Statistics` → `Summaries` → `Active data set`

2. Representing the data on a graph:
`Graphs` → `Scatterplot...` Then select variables x and y. Unfortunately, the only graphs available only represent the variables two by two.

3. Calculating the parameters:
`Statistics` → `Fit models` → `Linear regression...` then select the explanatory variables (x) and the response variable (y). It is then possible to obtain a confidence interval for each parameter of the regression:
`Models` → `Confidence intervals...`

4. Choosing the variables:
Many different models must be constructed and tested (the leaps package is not implemented): `Models` → `Hypothesis tests` → `Compare two models...`

5. Conducting a residual analysis:
`Models` → `Add observation statistics to data...`

We can add to the dataset the fitted values, the residuals, the studentised residuals, the predicted values, and Cook's distances, and create graphs with these new variables. Many graphs are available directly using
`Models` → `Graphs` → `Basic diagnostic plots`

6. Predicting a new value:
It is not possible to predict a new value automatically with Rcmdr.

7.2.6 Taking Things Further

Theoretical reminders and exercises on linear regression are available in many books, such as Clarke and Cooke (2004), Moore et al. (2007), Faraway (2005) or Fox and Weisberg (2011). Linear regression and many extensions are also available in Hastie et al. (2009).

7.3   Partial Least Squares (PLS) Regression

7.3.1   Objective

PLS regression consists of predicting a quantitative variable Y from p quantitative variables $X_1, \ldots, X_p$. Multiple regression (Section 7.2) is also dedicated to this purpose. However, when the number p of explanatory variables is very large, or if it exceeds the number n of individuals to be analysed, it is difficult, or even impossible, to use the least squares method. PLS regression is then particularly suited for cases with data with large dimensions such as spectrometry data or genetic data.

Put simply, PLS regression iteratively searches for a sequence of orthogonal components (or underlying variables). These components, known as PLS components, are chosen in order to maximise covariance with the explanatory variable Y. The choice of the number of components is important as it greatly influences the quality of the prediction. Indeed, if the chosen number is not sufficient, important information can be forgotten. However, if the chosen number is too big, the model may be overfitted. This choice is generally conducted using cross-validation or training/validation. These two procedures follow the same principle: The initial data is separated into two distinct parts, a training part and a validation part. There are generally more individuals in the training set than in the validation set. We construct the PLS regression model from the training set and then calculate the predictions on the validation set. We first construct the one-component PLS model and then repeat this procedure for $2, 3, \ldots, k$ PLS components. The number of components retained corresponds to the model which leads to the minimum prediction error. This procedure is implemented in the pls package.

7.3.2   Example

We would like to predict the organic carbon content in the ground from spectroscopic measurements. As measuring the organic carbon content of the ground is more expensive than collecting spectroscopic measurements, we would like to be able to predict organic carbon content from spectroscopic measurements.

In order to construct a prediction model, we have access to a database[1] with 177 soils (individuals) and both their organic carbon content (OC, response variable) and the spectra acquired in the visible and near-infrared range (400 nm–2500 nm), which gives 2101 explanatory variables. We will then predict the organic carbon content of three new soils.

---

[1]Thank to Y. Fouad and C. Walter; Aïchi H., Fouad Y., Walter C., Viscarra Rossel R.A., Lili Chabaane Z. and Sanaa M. (2009). Regional predictions of soil organic carbon content from spectral reflectance measurements. *Biosystems Engineering*, **104**, 442–446.

7.3.3 Steps

1. Read the data.

2. Represent the data.

3. Conduct a PLS regression after choosing the number of PLS components.

4. Conduct a residual analysis.

5. Predict a new value.

7.3.4 Processing the Example

1. Reading the data:

While being imported, an "X" is added by default to the names of variables that correspond to wavelength. Spectral data often needs to be preprocessed; this is usually done by standardising the data from a spectrum. To do so, we calculate the mean and the standard deviation for each spectrum (i.e. for each row in the matrix, omitting the first column which corresponds to Y). Then, for all the data in a row, we subtract its mean and divide by its standard deviation:

```
> Spectrum <- read.table("Spectrum_Breizh.txt", sep=";",
      header=TRUE, row.names=1)
> row.mean <- apply(Spectrum[,-1], 1, mean)
> Spectrum[,-1] <- sweep(Spectrum[,-1], 1, row.mean, FUN="-")
> row.sd <- apply(Spectrum[,-1], 1, sd)
> Spectrum[,-1] <- sweep(Spectrum[,-1], 1, row.sd, FUN="/")
```

2. Representing the data:

The first step is to visualise the response variable Y, which corresponds to organic carbon content OC, through the representation of its histogram:

```
> hist(Spectrum[,"OC"], freq=F, xlab="Organic Carbon content")
> lines(density(Spectrum[,"OC"]))
> rug(Spectrum[,"OC"])
```

The **rug** function is used to add vertical lines to the abscissa for the observations of variable Y. We can see in Figure 7.7 that one observation is extreme: It is greater than 8, whereas all the others are less than 5. The following lines can be used to show that individual 79 has a very high value (8.9) The other values do not exceed 4.89.

```
> which(Spectrum[,1]>8)
[1] 79
> Spectrum[79,1]
```

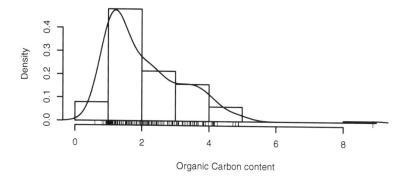

Figure 7.7
Representation of organic carbon content.

```
[1] 8.9
> max(Spectrum[-79,1])
[1] 4.89
```

This individual is unusual and can be removed when constructing the model:

```
> Spectrum <- Spectrum[-79,]
```

The explanatory variables are spectra and can be represented by curves. Each spectrum can be represented by a different line (continuous, dashed, etc.) and/or colour, depending on the value of the OC variable. To construct this graph, we divide the OC variable into seven classes of almost equal size using the **cut** function; the points at which they are cut are determined by the quantiles (**quantile**). The factor obtained in this way is converted to a colour and line type code using the **as.numeric** function and the arguments col and lty. On the abscissa axis, we use the wavelengths:

```
> color <- as.numeric(cut(Spectrum[,1], quantile(Spectrum[,1],
     prob=seq(0,1,by=1/7)), include.lowest = TRUE))
> matplot(x=400:2500, y=t(as.matrix(Spectrum[,-1])),type="l",
     lty=color,col=color,xlab="Wavelength",ylab="Reflectance")
```

We obtain the graph in Figure 7.8 in which each individual is a different curve.

It would seem possible to distinguish groups of curves drawn with the same code, which means that similar curves admit similar values for the OC variable.

3. Conducting a PLS regression after choosing the number of PLS components:

To conduct this regression, we use the preinstalled **pls** package that must be

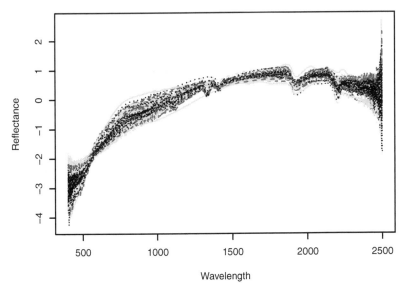

Figure 7.8
Representation of individuals (after a standardisation by row).

loaded (see Section 1.6). By default, this package centres all the variables but does not reduce them. We reduce the explanatory variables using the argument `scale`. We must also fix a maximum number of PLS components (the higher this number, the longer it will take to run the program). To be safe, we fix this number at 100, which is already extremely high.

```
> library(pls)
> pls.model <- plsr(OC~., ncomp=100, data=Spectrum,
    scale=TRUE, validation="CV")
```

With the pls package, the number of PLS components is, by default, determined by cross validation. We calculate the fitting error and the prediction error obtained with 1, 2, ..., 100 PLS components and we then draw the two types of error using the **plot.mvr** function (Figure 7.9):

```
> msepcv.pls <- MSEP(pls.model, estimate=c("train","CV"))
> plot(msepcv.pls,col=1,type="l",legendpos="topright",main="")
```

The two error curves are typical: a continual decline for the fitting error depending on the number of components and a decrease followed by an increase for the prediction error. The optimal number of components for the prediction corresponds to the value for which the prediction error is minimal:

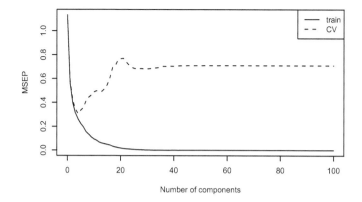

Figure 7.9
Evolution of both errors depending on the number of PLS components.

```
> ncp <- which.min(msepcv.pls$val["CV",,])-1
> ncp
4 comps
       4
```

We subtract 1 to obtain the number of PLS components directly (here there are four). As previously indicated, it was not necessary to fix the maximum number of PLS components at 100; we could have limited ourselves to 30. We fit the final model using the four PLS components:

```
> reg.pls <- plsr(OC~., ncomp=ncp, data=Spectrum, scale=TRUE)
```

4. Conducting a residual analysis:

The residuals are obtained using the **residuals** function. Note that the residuals are provided for the one-component PLS model, and then the two-component model, ..., until the model with the chosen number of components (here, four). We therefore only draw the residuals for the last model (Figure 7.10):

```
> res.pls <- residuals(reg.pls)
> plot(res.pls[,,ncp],pch=15,cex=.5,ylab="Residuals",main="")
> abline(h=c(-2,0,2), lty=c(2,1,2))
```

No individuals are so extreme that they need to be removed from the analysis. It must be noted that if individual 79 had not been deleted previously, it would have had a very large residual and would have been deleted at this point. The variability of residuals is greater for the last individuals of the dataset (individuals coded by `rmqs`) and smaller for the first (individuals coded by `Bt`). The samples from the units analysed do not seem very homogeneous.

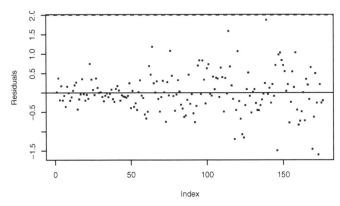

Figure 7.10
Representation of residuals.

5. Predicting a new value:

We have obtained a set of three complementary curves. For each one, we will compare the OC content predicted by the PLS regression to the observed values.

First we read the data and apply the same preprocessing step as before:

```
> ReflecN <- read.table("Spectrum_new.txt", sep=";", header=TRUE,
    row.names=1)
> row.meanN <- apply(ReflecN[,-1], 1, mean)
> ReflecN[,-1] <- sweep(ReflecN[,-1], 1, row.meanN, FUN="-")
> row.sdN <- apply(ReflecN[,-1], 1, sd)
> ReflecN[,-1] <- sweep(ReflecN[,-1], 1, row.sdN, FUN="/")
```

The predicted values are obtained using the **predict** function:

```
> pred <- predict(reg.pls, ncomp=ncp, newdata=ReflecN[,-1])
> pred
, , 4 comps

                OC
3236      0.4065104
rmqs_726 2.6930854
rmqs_549 2.6035296
```

These predicted values can be compared with the observed values:

```
> ReflecN[,1]
[1] 1.08 1.60 1.85
```

7.3.5 Rcmdr Corner

It is not possible to conduct PLS regression with `Rcmdr`.

7.3.6 Taking Things Further

PLS regression is often associated with specific graphs which can be used to analyse the model more precisely and to describe the analysed sample.

To better understand the role of each of the variables in the regression model, the coefficients can be drawn (Figure 7.11) for each of the wavelengths using the following commands:

```
> plot(reg.pls, plottype="coef", comps=1:ncp, main="",
    legendpos="topleft", xlab="Wavelength", labels=400:2500)
```

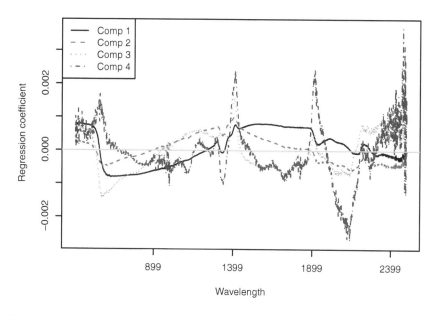

Figure 7.11
Coefficients of the PLS regression for one to four-component models.

The interpretation of this type of graph goes beyond the objectives set out in this work and we will simply underline the fact that there are coefficients for each reflectance of a given wavelength. They can therefore be interpreted in the same way as those of a multiple regression, the only difference being that here they are all on the same scale, as the initial variables (reflectances) are centred and reduced.

It is possible to represent the cloud of individuals on the PLS components (called "PLS scores") in order to visualise the similarities between individuals, as in PCA (see Worked Example 10.1). For example, we draw components 1 and 2 on an orthonormal index (`asp=1`) and colour all the samples from the `rmqs` group in grey using the following commands:

```
> color <- rep(1,nrow(Spectrum))
> color[substr(rownames(Spectrum),0,4)=="rmqs"] <- "grey30"
> plot(reg.pls, plottype="scores", comps=c(1,2),
      col=color, asp=1)
> abline(h=0,lty=2)
> abline(v=0,lty=2)
```

In Figure 7.12 we can see more clearly the division of the observed individuals into two populations.

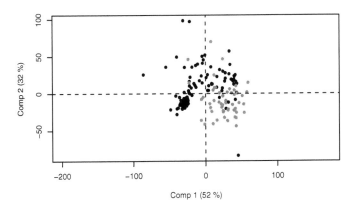

Figure 7.12
Analysis of the model by component: components 1 and 2.

Another classical analysis consists of representing the correlations between the loadings and the variables as in PCA (see graph of variables, Worked Example 10.1). We visualise the variables which are the most closely linked to each of the components. The wavelengths can also be coloured using greyscale (from black for the first variables to light grey for the last, see Figure 7.13).

```
> color <- grey(seq(0,.9,len=ncol(Spectrum)))
> plot(pls.model,plottype="cor",comps=1:2,col=color,pch=20)
```

Since the variables (here the wavelengths) are sorted, the evolution of their relationship with the PLS components can be represented using the argument `plottype="loadings"` (Figure 7.14):

```
> plot(pls.model,plottype="loadings",labels=400:2500,comps=1:ncp,
      legendpos="topright",xlab="Wavelength",ylab="Loadings")
> abline(h=0,lty=2)
```

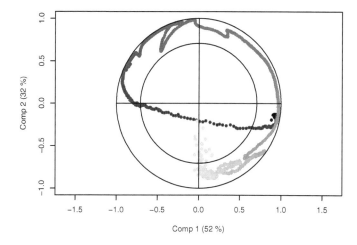

Figure 7.13
Representation of the correlations with the loadings.

A brief presentation of the PLS regression can be found in Hastie et al. (2009, p. 80–82). For more details and some case studies, see Varmuza and Filmozer (2009) or Vinzi et al. (2010).

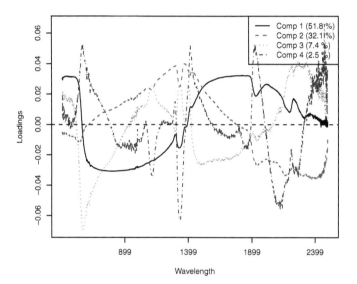

Figure 7.14
Representation of loadings.

8

Analysis of Variance and Covariance

8.1 One-Way Analysis of Variance

8.1.1 Objectives

One-way analysis of variance (or ANOVA) is a statistical method used to model the relationship between an explanatory qualitative variable A with I categories (or levels) and a quantitative response variable Y. The main objective of the one-way analysis of variance is to compare the empirical means of Y for I levels of A. The one-way analysis of variance can therefore be considered an extension of the two-way comparison of means (see Worked Example 6.3).

The one-way analysis of variance is therefore used to study the effect of a factor A on a quantitative variable Y. To do this, we construct the following model:

$$y_{ij} = \mu_i + \varepsilon_{ij}, \quad i = 1, ..., I \quad j = 1, ..., n_i$$

where n_i is the sample size of category i, y_{ij} the observed value j for sub-population i, μ_i the mean of sub-population i, and ε_{ij} the residual of the model. An individual is therefore here defined by the pair (i, j). The analysis of variance thus consists of testing the equality of the μ_i's.

The model can also be written in the following classical format:

$$y_{ij} = \mu + \alpha_i + \varepsilon_{ij}$$

where μ is the overall mean and α_i is the specific effect of category i. In this latest formulation, there are $I + 1$ parameters to be estimated, of which only I are identifiable. A linear constraint must therefore be imposed. There are a number of different constraints and the most common are

- One of the α_i is set at zero, which means that level i is considered the reference level.

- The sum of all α_i's is null, so the mean is taken as reference.

The numerical estimations of the parameters, obtained from the least squares method, depend of course on the chosen constraint. Nevertheless, whatever the constraint, it is always possible to test the overall significance of the factor. This test, which does not depend on the constraint, is Fisher's exact test. In this test, the variability explained by factor A (the between-group variability) is compared with the residual variability (the within-group variability). The hypotheses of this test are

$$H_0 : \quad \forall i \quad \alpha_i = 0 \quad \text{compared with} \quad H_1 : \exists i \quad \alpha_i \neq 0.$$

These hypotheses require us to test the sub-model with the complete model:

$$y_{ij} = \mu + \varepsilon_{ij}, \quad \text{model under } H_0$$
$$y_{ij} = \mu + \alpha_i + \varepsilon_{ij}, \quad \text{model under } H_1$$

Finally, a subsequent residual analysis is essential as it is used to check the individual fit of the model (outlier) and the overall fit.

Remark

In analyses of variance, the collected data is often rewritten in table format as in Table 8.1:

TABLE 8.1

Some Ozone Data Organized into Cells

Wind	Ozone Values					
North	87	82	114	79	101	...
South	90	72	146	108	81	...
East	92	121	146	106	45	...
West	94	80	79	106	70	...

In this table, the individuals are indexed conventionally in the format y_{ij} where i is the row index and j the column index. This presentation of the data indeed validates the determination of the model. Nevertheless, prior to the analysis, the data must be transformed by constructing a classic table of individuals × variables by considering one variable for the factor and another for the quantitative variable. For example, this yields Table 8.2.

8.1.2 Example

We reexamine in more detail the ozone dataset of Worked Example 7.1 (p. 133). Here we analyse the relationship between the maximum daily ozone concentration (in $\mu g/m^3$) and wind direction classed by sector (North, South, East, West). The variable wind has $I = 4$ levels. We have at our disposal 112 pieces of data collected during the summer of 2001 in Rennes (France).

8.1.3 Steps

1. Read the data.

TABLE 8.2

Presentation of the Data Table

Wind	Ozone
North	87
North	82
North	114
South	90
East	92
West	94
. . .	. . .

2. Represent the data.

3. Analyse the significance of the factor.

4. Conduct a residual analysis.

5. Interpret the coefficients.

8.1.4   Processing the Example

1. Reading the data:

Import the dataset and summarise the variables of interest, here `maxO3` and `wind`:

```
> ozone <- read.table("ozone.txt",header=T)
> summary(ozone[,c("maxO3","wind")])
     maxO3             wind
 Min.    : 42.00   East :10
 1st Qu.: 70.75   North:31
 Median : 81.50   South:21
 Mean    : 90.30   West :50
 3rd Qu.:106.00
 Max.    :166.00
```

During the summer of 2001, the most common wind direction was West, and there were very few days with an East wind.

2. Representing the data:

Prior to an analysis of variance, boxplots are usually constructed for each level of the qualitative variable. We therefore present the distribution of `maxO3` according to wind direction:

```
> plot(maxO3~wind,data=ozone,pch=15,cex=.5)
```

Looking at the graph in Figure 8.1, it would seem that there is a `wind` effect. This impression can be validated by testing the significance of the factor.

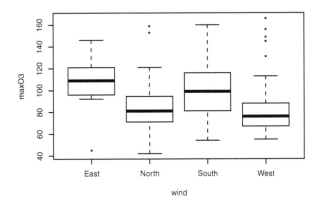

Figure 8.1
Boxplots of `maxO3` according to the levels of the variable `wind`.

3. Analysing the significance of the factor:
First of all, we use the **lm** (linear model) function (or the **aov** function) with a formula (see Appendix A.2, p. 266) in order to estimate the parameters of the model. The **anova** function returns the analysis of variance table:

```
> reg.aov1 <- lm(max03~wind,data=ozone)
> anova(reg.aov1)
Analysis of Variance Table

Response: max03
            Df Sum Sq Mean Sq F value  Pr(>F)
wind         3   7586 2528.69  3.3881 0.02074 *
Residuals  108  80606  746.35
---
Signif. codes:  0 '***' 0.001 '**' 0.01 '*' 0.05 '.' 0.1 ' ' 1
```

The first column indicates the factor's associated degrees of freedom, the second the sum of the squares, and the third the mean square (sum of the squares divided by the degrees of freedom). The fourth column features the observed value of the test statistic. The fifth column (`Pr(>F)`) contains the p-value, that is to say, the probability that the test statistic under H_0 will exceed the estimated value. The p-value (0.02) is less than 5%, thus H_0 is rejected and we can accept the significance of `wind` at the 5% level. There is therefore at least one wind direction for which the maximum level of ozone is significantly different from the others.

4. Conducting a residual analysis:

Residuals are available using the **residuals** function but they do not have the same variance (heteroscedastic). We therefore use studentised residuals. In order to plot the residuals according to the levels of the variable `wind`, we use the package lattice which is presented in detail in Section 3.2 (p. 73).

```
> res.aov1 <- rstudent(reg.aov1)
> library(lattice)
> mypanel <- function(...) {
    panel.xyplot(...)
    panel.abline(h=c(-2,0,2),lty=c(3,2,3),...)
  }
> trellis.par.set(list(fontsize=list(point=5,text=8)))
> xyplot(res.aov1~I(1:112)|wind,data=ozone,pch=20,
      ylim=c(-3,3),panel=mypanel, ylab="Residuals",xlab="")
```

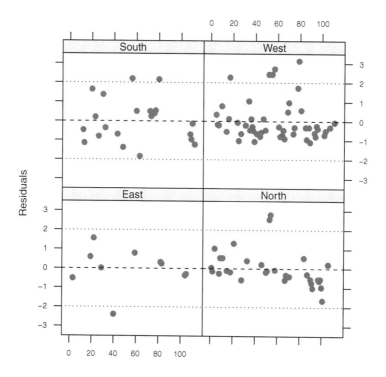

Figure 8.2

Representation of the residuals according to the levels of the variable `wind`.

In theory, 95% of the studentised residuals can be found in the interval $[-2, 2]$. Here there are nine residuals outside of the interval, that is, 8%, which is acceptable. The distribution of the residuals seems to be comparable from one category to another.

5. Interpreting the coefficients:
Now that an overall wind effect has been identified, we must examine how direction influences maximum ozone levels. In order to do so, the coefficients are analysed using the Student's t-test.

As mentioned above, there are a number of different ways of writing the one-way analysis of variance model. By default, R uses the constraint $\alpha_1 = 0$, which means taking as reference the first label of the variable alphabetically, which here is East.

```
> summary(reg.aov1)
Call:
lm(formula = maxO3 ~ wind, data = ozone)

Residuals:
    Min      1Q  Median      3Q     Max
-60.600 -16.807  -7.365  11.478  81.300

Coefficients:
            Estimate Std. Error t value Pr(>|t|)
(Intercept)  105.600      8.639  12.223   <2e-16 ***
windNorth    -19.471      9.935  -1.960   0.0526 .
windSouth     -3.076     10.496  -0.293   0.7700
windWest     -20.900      9.464  -2.208   0.0293 *
---
Signif. codes:  0 '***' 0.001 '**' 0.01 '*' 0.05 '.' 0.1 ' ' 1

Residual standard error: 27.32 on 108 degrees of freedom
Multiple R-Squared: 0.08602,    Adjusted R-squared: 0.06063
F-statistic: 3.388 on 3 and 108 DF,  p-value: 0.02074
```

Amongst other things, we obtain a `Coefficient` matrix which, for each parameter, (each line), has four columns: its estimation (`Estimate` column), its estimated standard deviation (`Std. Error`), the observed value of the test under $H_0 : \alpha_i = 0$ against $H_1 : \alpha_i \neq 0$. Finally, the associated p-value (`Pr(>|t|)`) yields the probability of exceeding the estimated value.

The estimate of μ, here denoted `Intercept`, is the mean East wind concentration in `maxO3`. The other values obtained correspond to the deviation from this mean for each wind cell.

The last three lines of column `Pr(>|t|)` correspond to the test $H_0: \alpha_i = 0$. The following question can thus be answered: Is there a similarity between the wind considered and the wind from reference cell East? The South wind is not different, unlike the West wind. The *p*-value associated with the North wind is slightly more than 5% and thus we cannot confirm a significant difference from the East wind.

If we want to select another specific control cell, for example the second label, which in this case is North, it can be indicated as follows:

```
> summary(lm(max03~C(wind,base=2),data=ozone))
Call:
lm(formula = max03 ~ C(wind, base = 2), data = ozone)

Residuals:
    Min      1Q  Median      3Q     Max
-60.600 -16.807  -7.365  11.478  81.300

Coefficients:
                    Estimate Std. Error t value Pr(>|t|)
(Intercept)           86.129      4.907  17.553   <2e-16 ***
C(wind, base = 2)1    19.471      9.935   1.960   0.0526 .
C(wind, base = 2)3    16.395      7.721   2.123   0.0360 *
C(wind, base = 2)4    -1.429      6.245  -0.229   0.8194
---
Signif. codes:  0 '***' 0.001 '**' 0.01 '*' 0.05 '.' 0.1 ' ' 1

Residual standard error: 27.32 on 108 degrees of freedom
Multiple R-Squared: 0.08602,    Adjusted R-squared: 0.06063
F-statistic: 3.388 on 3 and 108 DF,  p-value: 0.02074
```

Another possible constraint is $\sum_i \alpha_i = 0$, and is used as follows:

```
> summary(lm(max03~C(wind,sum),data=ozone))
Call:
lm(formula = max03 ~ C(wind, sum), data = ozone)

Residuals:
    Min      1Q  Median      3Q     Max
-60.600 -16.807  -7.365  11.478  81.300

Coefficients:
                 Estimate Std. Error t value Pr(>|t|)
(Intercept)        94.738      3.053  31.027   <2e-16 ***
C(wind, sum)1      10.862      6.829   1.590   0.1147
C(wind, sum)2      -8.609      4.622  -1.863   0.0652 .
C(wind, sum)3       7.786      5.205   1.496   0.1376
---
```

Signif. codes: 0 '***' 0.001 '**' 0.01 '*' 0.05 '.' 0.1 ' ' 1

Residual standard error: 27.32 on 108 degrees of freedom
Multiple R-Squared: 0.08602,   Adjusted R-squared: 0.06063
F-statistic: 3.388 on 3 and 108 DF, p-value: 0.02074

By default, R provides the values of $\hat{\mu}$, $\hat{\alpha}_1$, $\hat{\alpha}_2$ and $\hat{\alpha}_3$. As $\sum_i \alpha_i = 0$, in order to find the coefficient associated with the West wind (last level), we must calculate

$$\hat{\alpha}_4 = -\hat{\alpha}_1 - \hat{\alpha}_2 - \hat{\alpha}_3 = -10.038$$

In all these analyses, it can be seen that the values of the estimates change depending on the constraint. On the other hand, the overall test given in the last row of the lists, which corresponds to the result of the analysis of variable table, remains the same.

If within the same session, we want to carry out multiple analyses of variance with the same constraint, it is preferable to use

> **options**(contrasts = c("contr.sum", "contr.sum"))

the model is therefore written:

> **summary**(lm(maxO3~wind,data=ozone))

8.1.5   Rcmdr Corner

1. Reading the data from a file:
Data → Import data → from text file, clipboard, or URL ...

Next specify that the field separator is the space and the decimal-point character is ".".
To check that the dataset has been imported successfully:
Statistics → Summaries → Active data set

2. Representing the data:
Graphs → Boxplot... Select the response variable maxO3 and click Plot by groups... then select the variable wind. It is possible to identify specific points using the mouse.

3. Analysing the significance of the factor:
Statistics → Fit models → Linear model... then select the response variable (y) and the explanatory variable (x).

The outputs here are estimations of parameters μ and α_i with constraint $\alpha_1 = 0$. To change this constraint to $\sum_i \alpha_i = 0$, use: Tools → Options... → then in contrasts, replace contr.Treatment and contr.poly by contr.sum. Then click on Exit and Restart R Commander.

To build the analysis of variance table:
`Models → Hypothesis tests → ANOVA table...`

4. Conducting a residual analysis:
Many graphs are available for residual analysis, but not that of the lattice package. Nevertheless, we can obtain a few graphs using `Models → Graphs`

5. Interpreting the coefficients:
See the item 3 "Analysing the significance of the factor".

8.1.6   Taking Things Further

It is possible to construct multiple comparisons of means tests. In this case we use the **TukeyHSD** function. This function can also be used for a one-way analysis of variance, but only works for two-or-more factor analyses of variance when the dataset is balanced.

Theoretical presentations and exercises on analysis of variance are available in many books, such as Clarke and Cooke (2004), Moore et al. (2007), Sahai and Ageel (2000) or Faraway (2005).

8.2 Multi-Way Analysis of Variance with Interaction

8.2.1 Objective

Multi-way analysis of variance (or ANOVA) is the generalisation of one-way ANOVA (see Worked Example 8.1). Using this method, it is possible to model the relationship between a quantitative variable and many qualitative variables. We examine a two-variable case, but the generalisation to multiple variables is immediate. Let us consider two qualitative explanatory variables, denoted A and B, and a quantitative response variable, denoted Y. We denote I as the number of levels of the variable A and J as that of variable B.

By analysing variance, when we have multiple measurements for each confrontation of a level of A with a level of B, it is possible and often interesting to process the complete model using interaction. Interaction means examining the simultaneous effect of factors A and B on variable Y. The model is classically written as:

$$y_{ijk} = \mu + \alpha_i + \beta_j + \gamma_{ij} + \varepsilon_{ijk} \tag{8.1}$$

where μ is the overall mean, α_i is the effect due to the i-th level of the factor A, β_j is the effect due to the j-th level of the factor B, γ_{ij} is the interaction effect when factor A is at level i and factor B is at level j, and ε_{ijk} is the residual term.

It is therefore interesting to test the hypotheses of an effect of factors A and B, and the interaction AB. In order to do so, we construct Fisher's exact test to compare the explained variability with the residual variability. We advise that you first test the significance of the interaction. Indeed, if the interaction is significant, the two factors are influential via their interaction and it is therefore not necessary to test their respective influence via the main effect. The hypotheses of the interaction test are

$$(\mathrm{H_0})_{AB}: \quad \forall (i,j) \quad \gamma_{ij} = 0 \quad \text{compared with} \quad (\mathrm{H_1})_{AB}: \exists (i,j) \quad \gamma_{ij} \neq 0.$$

These hypotheses involve testing the sub-model against the complete model:

$$\begin{aligned} y_{ijk} &= \mu + \alpha_i + \beta_j + \varepsilon_{ijk}, \quad \text{model under} \quad (\mathrm{H_0})_{AB} \\ y_{ijk} &= \mu + \alpha_i + \beta_j + \gamma_{ij} + \varepsilon_{ijk}, \quad \text{model under} \quad (\mathrm{H_1})_{AB} \end{aligned} \tag{8.2}$$

This test, which identifies the overall influence of the interaction of the factors, is simply a test between two nested models.

We can then proceed in the same manner to test the effect of a single factor. For example, in order to test the effect of factor A, we test the sub-model without factor A against the model with factor A but without the interaction:

$$\begin{aligned} y_{ijk} &= \mu + \beta_j + \varepsilon_{ijk} \quad \text{model under} \quad (\mathrm{H}_0)_A \\ y_{ijk} &= \mu + \alpha_i + \beta_j + \varepsilon_{ijk} \quad \text{model under} \quad (\mathrm{H}_1)_A \end{aligned}$$

It may also be interesting to estimate the parameters of the model (μ, α_i, β_j, γ_{ij}). Nevertheless, as for the one-way analysis of variance, there are more parameters in the model than there are calculable parameters: $1 + I + J + IJ$ parameters to be estimated, whereas only IJ are calculable. It is therefore necessary to impose $1 + I + J$ linearly independent constraints in order to make the system reversible.

The classical constraints are

1. Cell-analysis constraints:

$$\mu = 0, \qquad \forall i, \quad \alpha_i = 0, \qquad \forall j, \quad \beta_j = 0$$

2. Cell reference constraints:

$$\alpha_1 = 0, \qquad \beta_1 = 0, \qquad \forall i, \quad \gamma_{i1} = 0, \qquad \forall j, \quad \gamma_{1j} = 0$$

3. Sum constraints:

$$\sum_i \alpha_i = 0, \qquad \sum_j \beta_j = 0, \qquad \forall i, \quad \sum_j \gamma_{ij} = 0, \qquad \forall j, \quad \sum_i \gamma_{ij} = 0$$

8.2.2 Example

We reexamine in more detail the ozone dataset of Worked Example 7.1 (p. 133). Here we will analyse the relationship between the maximum daily ozone concentration (in $\mu g/m^3$) and **wind** direction classed into sectors: **North, South, East, West** and precipitation classed into two categories: **Dry** and **Rainy**. We have at our disposal 112 pieces of data collected during the summer of 2001 in Rennes (France). Variable A admits $I = 4$ levels and variable B has $J = 2$ levels.

8.2.3 Steps

1. Read the data.

2. Represent the data.

3. Choose the model.

4. Interpret the coefficients.

8.2.4 Processing the Example

1. Reading the data:
Import the dataset and summarise the variables of interest, here `max03`, `wind` and `rain`:

```
> ozone <- read.table("ozone.txt",header=T)
> summary(ozone[,c("max03","wind","rain")])
      max03              wind          rain
 Min.   : 42.00   East :10    Dry  :69
 1st Qu.: 70.75   North:31    Rainy:43
 Median : 81.50   South:21
 Mean   : 90.30   West :50
 3rd Qu.:106.00
 Max.   :166.00
```

2. Representing the data:
A boxplot is presented for each cell (confrontation of a level of `wind` with a level of `rain`) which yields $IJ = 8$ boxplots (Figure 8.3):

```
> boxplot(max03~wind*rain,data=ozone)
```

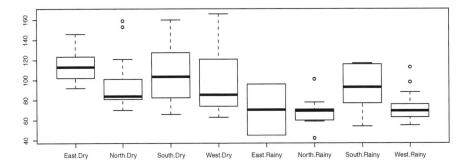

Figure 8.3
Boxplot of `max03` according to the confrontation of the levels of the variables `wind` and `rain`.

It must be noted that the ozone level is generally higher during dry weather than during rainy weather. Another way of graphically representing the interaction is as follows (see Figure 8.4):

```
> par(mfrow=c(1,2))
> with(ozone,interaction.plot(wind,rain,max03))
> with(ozone,interaction.plot(rain,wind,max03))
```

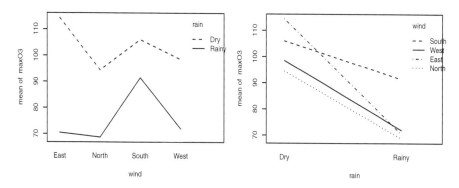

Figure 8.4
Interaction graph depicting wind (left) and rain (right) on the x-axis.

These graphs are interesting and can represent different interactions. We therefore advise that you construct both graphs even if only the more explicit of the two is analysed. The broken lines are almost parallel so the interaction is not very pronounced. We thus conduct the test to compare models (8.1) and (8.2).

3. Choosing the model:

```
> mod.int <- lm(maxO3~wind*rain,data=ozone)
> anova(mod.int)
Analysis of Variance Table

Response: maxO3
           Df Sum Sq Mean Sq F value  Pr(>F)
wind        3   7586  2528.7  4.1454    0.00809 **
rain        1  16159 16159.4 26.4910 1.257e-06 ***
wind:rain   3   1006   335.5  0.5500    0.64929
Residuals 104  63440   610.0
```

The first column indicates the factor's associated degrees of freedom, the second the sum of the squares, and the third the mean square (sum of the squares divided by the degrees of freedom). The fourth column features the observed value of the test statistic. The fifth column (`Pr(>F)`) contains the *p*-value, that is to say, the probability that the test statistic under H_0 will exceed the estimated value. The *p*-value (0.65) is more than 5%, thus we retain H_0 and conclude that the interaction is not significant. We therefore consider that the model has no interaction (8.2).

```
> mod.without.int <- lm(maxO3~wind+rain,data=ozone)
> anova(mod.without.int)
```

Analysis of Variance Table

```
Response: max03
            Df Sum Sq Mean Sq F value   Pr(>F)
wind         3   7586  2528.7  4.1984   0.007514 **
rain         1  16159 16159.4 26.8295 1.052e-06 ***
Residuals  107  64446   602.3
```

The two factors are significant and there is therefore both a wind effect and a rain effect on the maximum daily ozone level. We shall therefore retain this model and analyse the coefficients.

4. Interpreting the coefficients:
The calculations depend on the constraints used. By default, the constraints used by R are $\alpha_1 = 0$ and $\beta_1 = 0$. However, here there is no reason to fix a reference level. We therefore choose the constraints $\sum_i \alpha_i = 0$ and $\sum_j \beta_j = 0$, which yields

```
> summary(lm(max03~C(wind,sum)+C(rain,sum),data=ozone))
Call:
lm(formula = max03 ~ C(wind, sum) + C(rain, sum), data = ozone)

Residuals:
    Min      1Q  Median      3Q     Max
-42.618 -15.664  -3.712   8.295  67.990

Coefficients:
                Estimate Std. Error t value Pr(>|t|)
(Intercept)       90.135      2.883  31.260  < 2e-16 ***
C(wind, sum)1      7.786      6.164   1.263   0.2093
C(wind, sum)2     -8.547      4.152  -2.059   0.0420 *
C(wind, sum)3      5.685      4.694   1.211   0.2285
C(rain, sum)1     12.798      2.471   5.180 1.05e-06 ***
---
Signif. codes:  0 '***' 0.001 '**' 0.01 '*' 0.05 '.' 0.1 ' ' 1

Residual standard error: 24.54 on 107 degrees of freedom
Multiple R-Squared: 0.2692,     Adjusted R-squared: 0.2419
F-statistic: 9.856 on 4 and 107 DF,  p-value: 7.931e-07
```

Amongst other things, we obtain a `Coefficient` matrix with four columns for each parameter (row). These include its estimation (column `Estimate`), its estimated standard deviation (`Std. Error`), the observed value of the given test statistic, and finally the p-value (`Pr(>|t|)`) which yields under H_0 the probability of exceeding the estimated value.

The estimator of μ, here denoted `Intercept`, corresponds to the mean effect. The West wind effect is not included in the list. As the constraint is $\sum_i^I \alpha_i = 0$, we estimate α_4 for $-(7.786 - 8.547 + 5.685) = -4.924$. The effect of the category `Rainy` is -12.798. It is reassuring that the coefficient which corresponds to dry weather exceeds that associated with wet weather, as radiation is one of the catalysts of ozone production.

Like in one-way analysis of variance (see Worked Example 8.1), estimates depend on the constraints used. However, the overall test, which corresponds to the results of the analysis of variable table, remains the same.

If, within the same session, we want to carry out multiple analyses of variance with the same constraint, it is preferable to use

```
> options(contrasts = c("contr.sum", "contr.sum"))
```

The models are therefore written as:

```
> summary(lm(maxO3~wind+rain+wind:rain,data=ozone))
> summary(lm(maxO3~wind+rain,data=ozone))   #without interaction
```

8.2.5 Rcmdr Corner

1. Reading the data from a file:
`Data → Import data → from text file, clipboard, or URL ...`

Next specify that the field separator is the space and the decimal-point character is ".".
To check that the dataset has been imported successfully:
`Statistics → Summaries → Active data set`

2. Representing the data:
It is not possible to represent the interaction.

3. Choosing the model:
Multiple models can be chosen by using `Statistics → Fit models → Linear model...` and then writing the model. If multiple models have been constructed, it is possible to test one against another by using `Model → Hypothesis tests → Compare two models...`

By default, the outputs are estimates of parameters μ, α_i, β_j and γ_{ij} with "reference cell" constraints. To change this constraint to a sum constraint, use `Tools → Options... →` then in `contrasts`, replace `contr.Treatment` and `contr.poly` by `contr.sum`. Then click on `Exit and Restart R Commander`.

To draw up the analysis of variance table:
`Models → Hypothesis tests → ANOVA table`

4. Interpreting the coefficients:
See previous item.

8.2.6   Taking Things Further

During the construction of the model, we assume that the residuals follow a normal distribution and are homoscedastic. It can also be interesting to test the equality of variance of residuals for the different combinations of the interaction. In order to do so, we can construct a Bartlett test using the **bartlett.test** function. If we reject the equality of variance, we can construct a Friedman test, which is based on ranks, using the **friedman.test** function.

Theoretical presentations and exercises on analysis of variance are available in many books, such as Clarke and Cooke (2004), Faraway (2005), Moore et al. (2007) or Sahai and Ageel (2000).

8.3   Analysis of Covariance

8.3.1   Objective

The analysis of covariance aims to model the relationship between a quantitative response variable Y and quantitative and qualitative explanatory variables. This worked example is dedicated to examining the simplest case: that is, a case in which there are only two explanatory variables, one of which is quantitative and the other qualitative. The quantitative variable is denoted X and the qualitative variable Z, which is considered to have I categories. As the relationship between the response variable Y and the explanatory variable X can depend on the categories of the qualitative variable Z, the natural procedure involves conducting I different regressions, one for each level i of Z. In terms of modelling, this yields:

$$
\begin{aligned}
y_{1,j} &= \alpha_1 + \gamma_1 x_{1,j} + \varepsilon_{1,j} \quad j = 1, \ldots, n_1 \quad \text{level 1 of } Z \\
y_{2,j} &= \alpha_2 + \gamma_2 x_{2,j} + \varepsilon_{2,j} \quad j = 1, \ldots, n_2 \quad \text{level 2 of } Z \\
&\vdots \qquad\qquad \vdots \\
y_{I,j} &= \alpha_I + \gamma_I x_{I,j} + \varepsilon_{I,j} \quad j = 1, \ldots, n_I \quad \text{final level of } Z
\end{aligned}
$$

Put more simply

$$
y_{i,j} = \alpha_i + \gamma_i x_{i,j} + \varepsilon_{i,j} \quad i = 1, \ldots, I, \; j = 1, \ldots, n_i \qquad (8.3)
$$

In this model the $2I$ parameters (α_i and γ_i) are calculated by the least squares method by conducting a simple regression for each level of Z. In terms of error, one typically considers that all the $\varepsilon_{i,j}$ have the same variance σ^2, an assumption made both in regression and analysis of variance.

This model can also be written by identifying an overall mean α and a mean slope γ (as in analysis of variance):

$$
y_{i,j} = \alpha + \alpha_i + (\gamma + \gamma_i) x_{i,j} + \varepsilon_{i,j} \quad i = 1, \ldots, I, \; j = 1, \ldots, n_i. \;(8.4)
$$

Constraints must thus be added in order to obtain an identifiable model. As in analysis of variance, we classically impose $\sum_i \alpha_i = 0$ and $\sum_i \gamma_i = 0$, or indeed one $\alpha_i = 0$ and one $\gamma_i = 0$.

To know whether Z has an effect on Y, we need to test the equality of the intercepts and that of the slopes between different models of linear regression. A number of different cases could be considered (see Figure 8.5).

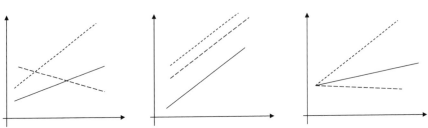

Figure 8.5
Illustration of the models (8.4), (8.5) and (8.6).

The more general model (8.4), which we refer to as the complete model, is illustrated on the left in Figure 8.5. In this case, the slopes and intercepts are assumed to be different for each level of Z.

A first simplification of this model is to suppose that variable X intervenes in the same manner, whatever the level of variable Z. It amounts to assuming that the slopes are identical, but that the intercepts are not. This is the model illustrated in the centre of Figure 8.5, which is written as

$$y_{i,j} \;=\; \alpha_i + \gamma x_{i,j} + \varepsilon_{i,j} \quad i = 1, \ldots, I, \quad j = 1, \ldots, n_i \qquad (8.5)$$

It must be noted that here there is no interaction between X and Z as the slopes are identical.

Another simplification of the complete model (8.4) is to assume that in the categories of Z, the intercepts are the same but the slopes are not (see right-hand graph in Figure 8.5). The model is therefore written as

$$y_{i,j} \;=\; \alpha + \gamma_i x_{i,j} + \varepsilon_{i,j} \quad i = 1, \ldots, I, \quad j = 1, \ldots, n_i \qquad (8.6)$$

The choice of model (8.4), (8.5) or (8.6) depends on the problem addressed. We advise starting with the most general model (8.4) and then, if the slopes are the same, we turn to model (8.5); if the intercepts are the same, we turn to model (8.6). As the models are nested, it is possible to test one model against another.

Once the model is chosen, the residuals must be analysed. This analysis is essential as it is used to check the individual fit of the model (outliers) and the global fit, for example by checking that there is no structure.

8.3.2 Example

We are interested in the balance of flavours between different ciders and, more precisely, the relationship between the sweet and bitter flavours according to the type of cider (dry, semi-dry or sweet). The flavour evaluations were provided by a tasting panel (mean marks, from 1 to 10, by 24 panel members) for each of 50 dry ciders, 30 semi-dry ciders and 10 sweet ciders.

8.3.3 Steps

1. Read the data.

2. Represent the data.

3. Choose the model.

4. Conduct a residual analysis.

8.3.4 Processing the Example

1. Reading the data:
Import the data, check the number of samples and the number of variables in the dataset, and summarise the variables of interest, here Type, Sweetness and Bitterness:

```
> cider <- read.table("cider.csv",header=TRUE,sep=",")
> dim(cider)
[1] 90  5
> summary(cider[,c(1,2,4)])
      Type         Sweetness            Bitterness
 Dry     :50   Min.    :3.444    Min.    :2.143
 Semi-dry:30   1st Qu.:4.580     1st Qu.:3.286
 Sweet   :10   Median :5.250     Median :3.964
               Mean    :5.169    Mean    :4.274
               3rd Qu.:5.670     3rd Qu.:5.268
               Max.    :7.036    Max.    :7.857
```

2. Representing the data:
Prior to an analysis of covariance, it may be useful to explore the data by representing, for each level of the qualitative variable, the scatterplot confronting the response variable Sweetness and the quantitative explanatory variable Bitterness. It is possible to represent these scatterplots on the same graph using different symbols (Figure 8.6). To do so, we use the argument pch and convert the variable Type as numeric to define pch values from 1 to the number of levels according to the values of Type. The argument col can be used exactly in the same way if we want to use a specific colour for each cider. Then, we add a legend on the top right of the graph to specify the symbol used for each type of cider (Figure 8.6).

```
> plot(Sweetness~Bitterness,pch=as.numeric(Type),data=cider)
> legend("topright",levels(cider$Type),pch=1:nlevels(cider$Type))
```

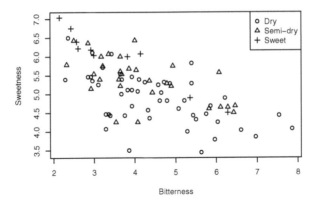

Figure 8.6
Representation of the scatterplot.

It is also possible to represent the data using the **xyplot** function from the lattice package (see Figure 8.7):

```
> library(lattice)
> xyplot(Sweetness~Bitterness|Type,data=cider)
```

To have a more precise idea of the effect of the qualitative variable, we could represent the regression line for each category but it is difficult to get an idea of the most appropriate model simply by looking at these graphs. It is therefore preferable to analyse the models.

3. Choosing the model:
The complete model defined by (8.4) is adjusted using the **lm** (linear model) function:

```
> global <- lm(Sweetness~-1+Type+Type:Bitterness, data=cider)
```

We remove the intercept by writing −1 but it must be specified that it is necessary to calculate an intercept for each type of cider by adding the factor Type. We must then specify that we require a different slope for each type of cider by writing the interaction of the variables Sweetness and Type.

The one-slope model (8.5) can be written as

```
> slopeU <- lm(Sweetness~-1 + Type + Bitterness, data = cider)
```

and a model with only one intercept (8.6)

```
> interceptU <- lm(Sweetness ~ Type:Bitterness, data = cider)
```

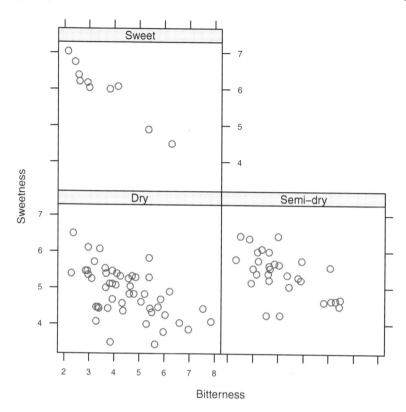

Figure 8.7
Scatterplots according to the categories of the variable `Type`.

In order to choose the best model, we can conduct multiple tests:

(a) slope or intercept equality: We conduct a test between the models (8.5) and (8.4) and another between the models (8.6) and (8.4) using the following commands:

```
> anova(slopeU,global)
Analysis of Variance Table

Model 1: Sweetness ~ -1 + Type + Bitterness
Model 2: Sweetness ~ -1 + Type + Type:Bitterness
  Res.Df    RSS Df Sum of Sq      F Pr(>F)
1     86 24.019
2     84 23.118  2   0.90126 1.6374 0.2006

> anova(interceptU,global)
```

```
Analysis of Variance Table

Model 1: Sweetness ~ Type:Bitterness
Model 2: Sweetness ~ -1 + Type + Type:Bitterness
  Res.Df      RSS Df Sum of Sq        F Pr(>F)
1      86 25.861
2      84 23.118  2    2.7428 4.983 0.009015 **
```

As the p-value (0.20) of the first test of nested models is greater than 5%, we can conclude that the slopes can be considered equal. As the p-value of the second test of nested models is less than 5%, we conclude that the intercepts are not equal. We therefore conserve the model without the interaction, which corresponds to model (8.5). This means that whatever the type of cider, the effect of the perception of bitterness and the perception of sweetness is the same. Now that we are in the (8.5) configuration, we will test whether or not it is possible to further simplify the model.

(b) null slope or intercept equality: We conduct a test between the model (8.5) and the simple regression model (see Worked Example 7.1): This test will reveal whether it is preferable to conserve the model with different intercepts, or if a simple linear regression model between the two quantitative variables is sufficient. We will then carry out a test between the model (8.5) and the one-way analysis of variance (see Worked Example 8.1) to know whether the latter is sufficient.

```
> simple <- lm(Sweetness ~ Bitterness,data = cider)
> anova(simple,slopeU)
Analysis of Variance Table

Model 1: Sweetness ~ Bitterness
Model 2: Sweetness ~ -1 + Type + Bitterness
  Res.Df      RSS Df Sum of Sq        F Pr(>F)
1      88 30.945
2      86 24.019  2    6.9256 12.399 1.857e-05 ***

> anova1 <- lm(Sweetness ~ Type,data = cider)
> anova(anova1,slopeU)
Analysis of Variance Table

Model 1: Sweetness ~ Type
Model 2: Sweetness ~ -1 + Type + Bitterness
  Res.Df      RSS Df Sum of Sq        F Pr(>F)
1      87 38.571
2      86 24.019  1   14.552 52.102 1.960e-10 ***
```

The two simplified models are rejected and the (8.5) model retained. The perception of bitterness thus has an influence on the perception of sweetness and this influence is the same for all types of cider (the slope is the same but the intercepts are different).

4. Conducting a residual analysis:
The residual analysis is similar to that conducted in Worked Example 7.1. Depending on the model chosen, we analyse the residuals using

```
> xyplot(rstudent(chosen_model)~Bitterness|Type, ylab="res",
    data=cider)
```

8.3.5   Rcmdr Corner

1. Reading the data from a file:
Data → Import data → from text file, clipboard, or URL ...

Next specify that the field separator is a comma.
To check that the dataset has been imported successfully:
Statistics → Summaries → Active data set

2. Representing the data on a graph:
Graphs → Scatterplot... Then select variables x and y. Then click on Plot by groups and choose the qualitative variable. This automatically produces a graph with a default colour for each level of the qualitative variable and the corresponding regression lines.

3. Choosing the model:
Multiple models can be chosen by using Statistics → Fit models → Linear model... and then writing the model. If multiple models have been constructed, it is possible to test one against another using Models → Hypothesis tests → Compare two models...

4. Conducting a residual analysis:
Models → Add observation statistics to data...

A number of graphs are made directly available. To construct them, use

Models → Graphs → Basic diagnostic plots

It is not possible to make the graph offered by lattice.

8.3.6   Taking Things Further

It is possible to construct an analysis of covariance model with all possible effects, and to test all the effects by decomposing the variability. This approach is easier when there are multiple qualitative and/or quantitative variables. The car package must be used in order to obtain a type III sum of squares decomposition of variation (in this way the same sum squares are obtained for any order of the effects in the model). The **Anova** function (with a capital A) is used to obtain the analysis of variance table. We therefore write

```
> library(car)
> global <- lm(Sweetness~Type+Bitterness+Type:Bitterness,
    data=cider)
> Anova(global, type="III")
Anova Table (Type III tests)

Response: Sweetness
                Sum Sq Df  F value      Pr(>F)
(Intercept)    144.139  1 523.7375 < 2.2e-16 ***
Type             2.743  2   4.9830  0.009015 **
Bitterness       6.879  1  24.9968 3.098e-06 ***
Type:Bitterness  0.901  2   1.6374  0.200634
Residuals       23.118 84
```

From these results, the effects with the highest *p*-value, and those over 5% can be eliminated.

Once the model has been chosen, it is possible to estimate the coefficients of the model using the **summary** function:

```
> global <- lm(Sweetness~Type+Bitterness,data=cider)
> summary(global)
Coefficients:
              Estimate Std. Error t value Pr(>|t|)
(Intercept)    6.30248    0.21230  29.687  < 2e-16 ***
TypeSemi-dry   0.41723    0.12293   3.394  0.00104 **
TypeSweet      0.83126    0.18797   4.422 2.84e-05 ***
Bitterness    -0.31928    0.04423  -7.218 1.96e-10 ***
```

The coefficients are estimated using the constraint that the first coefficient (that of the category Dry) is equal to 0. If we want to use the constraint $\sum_i \alpha_i = 0$, we use **C**(Type,sum) for the analysis of variance.

Theoretical presentations and exercises on analysis of covariance are available in many books, such as Faraway (2005).

9

Classification

9.1 Linear Discriminant Analysis

9.1.1 Objective

The aim of linear discriminant analysis is to explain and predict the values of a qualitative variable Y from quantitative and/or qualitative explanatory variables $X = (X_1, \ldots, X_p)$. We here suppose that the variable Y is binary, and the generalization to more than two categories is tackled towards the end of the worked example.

Linear discriminant analysis (LDA) can be presented according to two very different but equivalent aspects. Firstly, one point of view consists of modelling the probability of belonging to a group using Bayes' formula. To fix the notations, the categories of the response variable Y are designated 0 and 1. Bayes' rule estimates the posterior probability of allocation:

$$\mathbb{P}(Y = 1 | X = x) = \frac{\pi_1 f_1(x)}{\pi_0 f_0(x) + \pi_1 f_1(x)} \tag{9.1}$$

where $f_0(x)$ and $f_1(x)$ stand for the densities of X knowing that $Y = 0$ and $Y = 1$, and $\pi_0 = \mathbb{P}(Y = 0)$ and $\pi_1 = \mathbb{P}(Y = 1)$ represent the prior probabilities of belonging to classes 0 and 1, respectively. These two probabilities must be fixed by the user or estimated from the sample. In order to calculate the posterior probabilities $\mathbb{P}(Y = 0 | X = x)$ and $\mathbb{P}(Y = 1 | X = x)$, the discriminant analysis models the distributions of X knowing that $Y = j$ ($j = 0, 1$) by normal distributions. More specifically, we make the assumption

$$X | Y = 0 \sim \mathcal{N}(\mu_0, \Sigma), \qquad X | Y = 1 \sim \mathcal{N}(\mu_1, \Sigma).$$

The parameters μ_0, μ_1 and Σ are estimated by the method of maximum likelihood. For a new individual, we deduce from (9.1) the posterior probabilities and allocate this new individual to the group with the greatest one.

LDA can also be seen as a dimensionality reduction technique. In this case, as in principal component analysis, the principle consists of calculating a new variable, known as the canonical variable $w'X = w_1 X_1 + \ldots + w_p X_p$, as a linear combination of the initial variables. This variable is calculated in such a way as to maximise the ratio of the between-group and within-group variances (for further details, see McLachlan (2004), Chapter 18). For an individual $x = (x_1, \ldots, x_p)$, the canonical variable defines a score function $S(x) = w_1 x_1 + \ldots + w_p x_p$. An individual x is allocated to a given group by comparing the value of the score $S(x)$ with a threshold value s.

Remark

The two ways of presenting LDA do not allow to handle qualitative explanatory variables. However, as we shall see in the example, a disjunctive encoding of the explanatory variables can be used to conduct a linear discriminant analysis in the presence of such variables. Each category of the variable is therefore processed as a quantitative variable with values 0 or 1. In order to avoid problems of collinearity between the different categories, the column which is associated with the factor's first category is not taken into account in the model. When a discriminant analysis is conducted with one or more qualitative explanatory variables, the assumption of normality is clearly not verified: It is therefore important to be wary when interpreting posterior probabilities.

9.1.2   Example

We would like to explain the propensity to snore from six explanatory variables. The file snore.txt contains a sample of 100 patients. The variables considered in the sample are

- age: of the individual in years

- weight: of the individual in kilograms (kg)

- size: height of the individual in centimeters (cm)

- alcohol: number of units drunk per day (each equal to one glass of red wine)

- sex: of the individual (W = woman, M = male)

- snore: propensity to snore (Y = snores, N = does not snore)

- tobacco: smoking behaviour (Y = smoker, N = non-smoker)

The aim of this study is to try to explain snoring (variable snore) using the six other variables presented above. The following table presents an extract from the dataset:

	age	weight	size	alcohol	sex	snore	tobacco
1	47	71	158	0	M	N	Y
2	56	58	164	7	M	Y	N
⋮	⋮	⋮	⋮	⋮	⋮	⋮	⋮
99	68	108	194	0	W	Y	N
100	50	109	195	8	M	Y	Y

9.1.3   Steps

1. Read the data.

2. Construct the model.

3. Estimate the classification error rate.

4. Make a prediction.

9.1.4   Processing the Example

1. Reading the data:

```
> snore_data <- read.table("snore.txt",header=T)
```

Summarise the dataset:

```
> summary(snore_data)
      age              weight             size            alcohol
 Min.   :23.00    Min.   : 42.00    Min.   :158.0    Min.   : 0.00
 1st Qu.:43.00    1st Qu.: 77.00    1st Qu.:166.0    1st Qu.: 0.00
 Median :52.00    Median : 95.00    Median :186.0    Median : 2.00
 Mean   :52.27    Mean   : 90.41    Mean   :181.1    Mean   : 2.95
 3rd Qu.:62.25    3rd Qu.:107.00    3rd Qu.:194.0    3rd Qu.: 4.25
 Max.   :74.00    Max.   :120.00    Max.   :208.0    Max.   :15.00

 sex       snore      tobacco
 M:75      N:65       N:36
 W:25      Y:35       Y:64
```

2. Constructing the model:

The **lda** function from the MASS package is used to conduct the LDA. The MASS package is installed by default in R, and simply needs to be loaded. The **lda** function uses a formula like the function **lm** (see Appendix A.2, p. 266):

$$Y \; \tilde{} \; X_1 + X_2.$$

The prior probabilities also need to be specified. If there are no prior assumptions about these two quantities, there are two possible strategies:

- The probabilities are chosen to be equal, that is, $\pi_0 = \pi_1 = 0.5$;
- The probabilities are estimated by the proportion of observations in each group.

The choice of prior probabilities is specified by the `prior` argument of the **lda** function. If nothing is specified, the second strategy is chosen by default. We first write the model which takes all the explanatory variables into account:

```
> library(MASS)
> model <- lda(snore~.,data=snore_data)
> model

Call:
lda(snore ~ ., data = snore_data)

Prior probabilities of groups:
   N    Y
0.65 0.35

Group means:
       age    weight     size  alcohol       sexW   tobaccoY
N 50.26154 90.47692 180.9538 2.369231 0.3076923 0.6769231
Y 56.00000 90.28571 181.3714 4.028571 0.1428571 0.5714286

Coefficients of linear discriminants:
                 LD1
age         0.05973655
weight     -0.01620579
size        0.01590170
alcohol     0.24058822
sexW       -0.55413371
tobaccoY   -1.14621434
```

In the output, we find the prior probabilities of the model, the centre of gravity for each of the two groups, and the coefficients of the canonical variable. As a rule, the software returns the coefficients of the canonical variable so that the within-class variance equals 1. It is difficult to find the important variables for the discrimination simply by comparing the centres of gravity; it is often more relevant to study the influence of the coefficients on the score. For example, it can be seen that, when all the other variables are equal, a man will have a score 0.5541 higher than a woman. In the same way, a ten-year age difference between two individuals who otherwise share the same characteristics will contribute to an evolution in the score of 0.5974.

No statistical test can be used to test the significance of the coefficients of the canonical variable. Nevertheless here we can see that the variables weight and size are close to 0. It may therefore be interesting to delete these two variables in the model as follows:

```
> model1 <- lda(snore~age+alcohol+sex+tobacco,data=snore_data)
> model1
Call:
lda(snore ~ age + alcohol + sex + tobacco, data = snore_data)
```

```
Prior probabilities of groups:
   N    Y
0.65 0.35
```

```
Group means:
        age  alcohol       sexW  tobaccoY
N 50.26154 2.369231 0.3076923 0.6769231
Y 56.00000 4.028571 0.1428571 0.5714286
```

```
Coefficients of linear discriminants:
                  LD1
age         0.06062024
alcohol     0.24060629
sexW       -0.54047702
tobaccoY   -1.13039555
```

It can be seen that the coefficients of the canonical variable are only slightly affected by the removal of these two variables.

To impose equal prior probabilities on the model, we use the `prior` argument:

```
> model2 <- lda(snore~.,data=snore_data,prior=c(0.5,0.5))
> model3 <- lda(snore~age+alcohol+sex+tobacco,data=snore_data,
      prior=c(0.5,0.5))
> model2
Call:
lda(snore ~ ., data = snore_data, prior = c(0.5, 0.5))
```

```
Prior probabilities of groups:
  N   Y
0.5 0.5
```

```
Group means:
        age   weight     size  alcohol       sexW  tobaccoY
N 50.26154 90.47692 180.9538 2.369231 0.3076923 0.6769231
Y 56.00000 90.28571 181.3714 4.028571 0.1428571 0.5714286
```

```
Coefficients of linear discriminants:
                  LD1
age         0.05973655
weight     -0.01620579
size        0.01590170
alcohol     0.24058822
sexW       -0.55413371
tobaccoY   -1.14621434
```

The modification of prior probabilities does not change the canonical variable.

This choice only influences the way new individuals are allocated to a given group.

3. Estimating the classification error rate:

The **lda** function is used to estimate the classification error rate by cross validation. In order to do so, simply add the argument CV=TRUE to the function call. We obtain the labels predicted by the model named model using the following command:

```
> pred <- lda(snore~.,data=snore_data,CV=TRUE)$class
> table(pred,snore_data$snore)


pred  N  Y
   N 53 22
   Y 12 13
```

We thus obtain an estimation of the classification error rate:

```
> sum(pred!=snore_data$snore)/nrow(snore_data)
[1] 0.34
```

The classification error rate is estimated in the same way for the other models studied:

```
> pred1 <- lda(snore~age+alcohol+sex+tobacco,data=snore_data,
                CV=TRUE)$class
> table(pred1,snore_data$snore)
pred1  N  Y
    N 52 23
    Y 13 12
> sum(pred1!=snore_data$snore)/nrow(snore_data)
[1] 0.36

> pred2 <- lda(snore~.,data=snore_data,prior=c(0.5,0.5),
                CV=TRUE)$class
> table(pred2,snore_data$snore)
pred2  N  Y
    N 41 10
    Y 24 25
> sum(pred2!=snore_data$snore)/nrow(snore_data)
[1] 0.34

> pred3 <- lda(snore~age+alcohol+sex+tobacco,data=snore_data,
                prior=c(0.5,0.5),CV=TRUE)$class
> table(pred3,snore_data$snore)
```

```
pred3  N  Y
    N 42  9
    Y 23 26
> sum(pred3!=snore_data$snore)/nrow(snore_data)
[1] 0.32
```

The last model is the most effective in terms of misclassification. Furthermore, we can see that the last two models (those for which the prior probabilities are equal to 0.5) are also able to identify individuals who snore as they detect 25 and 26 out of 35, compared with 13 and 12 out of 35 for the first two models.

4. Making a prediction:

The models constructed previously can be used for prediction. The table below contains the values for six explanatory variables measured on four new patients:

Age	Weight	Size	Alcohol	Sex	Tobacco
42	55	169	0	W	N
58	94	185	4	M	Y
35	70	180	6	M	Y
67	63	166	3	W	N

In order to predict the labels of these four new individuals, we first collect the new data in a data-frame with the same structure as the initial data table (and in particular the same names for variables). We thus construct a new matrix for the quantitative variables and another for the qualitative variables before grouping them together in the same data-frame:

```
> new_d1 <- matrix(c(42,55,169,0,58,94,185,4,35,70,180,6,67,63,
    166,3),ncol=4,byrow=T)
> new_d2 <- matrix(c("W","N","M","Y","M","Y","W","N"),ncol=2,
    byrow=T)
> new_d <- cbind.data.frame(new_d1,new_d2)
> names(new_d) <- names(snore_data)[-6]
```

The **predict** function can be used to allocate each of these four individuals a group:

```
> predict(model,newdata=new_d)
$class
[1] N N N Y
Levels: N   Y
```

```
$posterior
          N         Y
1 0.7957279 0.2042721
2 0.6095636 0.3904364
3 0.7325932 0.2674068
4 0.3532721 0.6467279

$x
          LD1
1 -0.6238163
2  0.3246422
3 -0.2586916
4  1.4140108
```

The **predict** function returns a list of three objects:

- The first element (`class`) contains the predicted groups.
- The second element (`posterior`) is a two-column matrix featuring the posterior possibilities of belonging to groups N and Y for each individual.
- The third element (`x`) designates the score value for each individual.

The model allocates the first three individuals to group N and the fourth to Y. The rule for allocation to a group with an observation x is conducted by comparing the posterior probabilities or, in the same way, comparing the score with a threshold value. Modifying the prior probabilities does not change the canonical variable; the only thing that will change is the threshold from which we change the allocation to a group. Since the software does not return this threshold value, the allocations are made by comparing the posterior probabilities for each group.

9.1.5   Rcmdr Corner

It is not possible to conduct a linear discriminant analysis with `Rcmdr`.

9.1.6   Taking Things Further

To conduct an LDA, we assume that the variance-covariance matrices of the explanatory variables are the same for each group. This hypothesis can be avoided thanks to quadratic discriminant analysis. Posterior probabilities are calculated supposing that

$$X|Y = 0 \sim \mathcal{N}(\mu_0, \Sigma_0), \qquad X|Y = 1 \sim \mathcal{N}(\mu_1, \Sigma_1)$$

The **qda** function, available in the MASS package, can be used to conduct a quadratic discriminant analysis in R. It is used in the same way as **lda**.

In this chapter, we limited our analysis to a binary response variable. LDA can easily be adapted to cases where the response variable carries $J > 2$ categories. The posterior probabilities are still calculated using Bayes' formula by assuming normality for each group $X|Y = j \sim \mathcal{N}(\mu_j, \Sigma)$ for $j = 1, \ldots, J$ (or $X|Y = j \sim \mathcal{N}(\mu_j, \Sigma_j)$ for the quadratic discriminant). In this case, we are no longer looking for one canonical variable, but rather $(J-1)$ variables. The calculations can be implemented in R with no difficulty and are conducted in the same way: the models are written in the same way and individuals are allocated to the group with the highest posterior probability.

For further theoretical elements on linear discriminant analysis, refer to McLachlan (2004).

9.2 Logistic Regression

9.2.1 Objective

The aim of logistic regression is to explain and predict the values of a qualitative variable Y, which is generally binary, from qualitative and quantitative explanatory variables $X = (X_1, \ldots, X_p)$. If the categories of Y are denoted 0 and 1, the logistic model is expressed as

$$\log\left(\frac{p(x)}{1 - p(x)}\right) = \beta_0 + \beta_1 x_1 + \ldots + \beta_p x_p$$

where $p(x)$ denotes the probability $\mathbb{P}(Y = 1 | X = x)$ and $x = (x_1, \ldots, x_p)$ is a realization of $X = (X_1, \ldots, X_p)$. The coefficients $\beta_0, \beta_1, \ldots, \beta_p$ are estimated by the method of maximum likelihood.

9.2.2 Example

Treatment for prostate cancer changes depending on whether or not the lymphatic nodes surrounding the prostate are affected. In order to avoid an invasive investigative procedure (opening the abdominal cavity), a certain number of variables are considered explanatory variables for the binary variable Y: if $Y = 0$, the cancer has not reached the lymphatic system; if $Y = 1$, the cancer has reached the lymphatic system. The aim of this study is therefore to explain and predict Y from the following variables:

- Age of the patient at the time of diagnosis (age)

- Serum acid phosphatase level (acid)

- Result of an X-ray analysis, 0 = negative, 1 = positive (Xray)

- Size of the tumour, 0 = small, 1 = large (size)

- State of the tumour as determined by a biopsy, 0 = medium, 1 = serious (grade)

- Logarithm of the acidity level (log.acid)

The following table presents an extract from the dataset:

	age	acid	Xray	size	grade	Y	log.acid
1	66	0.48	0	0	0	0	-0.7339692
2	68	0.56	0	0	0	0	-0.5798185
⋮	⋮	⋮	⋮	⋮	⋮	⋮	⋮
52	64	0.89	1	1	0	1	-0.1165338
53	68	1.26	1	1	1	1	0.2311117

9.2.3 Steps

1. Read the data.

2. Construct the model.

3. Choose the model.

4. Make a prediction.

9.2.4 Processing the Example

1. Reading the data:
Reading the data from the file `prostate.txt`:

```
> prostate <- read.table("prostate.txt",header=T)
```

The variables `Xray`, `size`, `grade` and `Y` must be converted into factors:

```
> for (i in 3:6){prostate[,i] <- factor(prostate[,i])}
```

Summarise the dataset:

```
> summary(prostate)
```

```
      age              acid         Xray   size   grade  Y          log.acid
 Min.   :45.00   Min.   :0.4000   0:38   0:26   0:33   0:33   Min.   :-0.9163
 1st Qu.:56.00   1st Qu.:0.5000   1:15   1:27   1:20   1:20   1st Qu.:-0.6931
 Median :60.00   Median :0.6500                              Median :-0.4308
 Mean   :59.38   Mean   :0.6942                              Mean   :-0.4189
 3rd Qu.:65.00   3rd Qu.:0.7800                              3rd Qu.:-0.2485
 Max.   :68.00   Max.   :1.8700                              Max.   : 0.6259
```

2. Constructing the model:
In order to better explain the idea of logistic regression, we first consider the elementary logistic model used to explain variable Y by variable `log.acid`. Let us denote X the variable `log.acid` and x an occurrence of this variable. In this case, the logistic model is simply written as

$$\log\left(\frac{p(x)}{1-p(x)}\right) = \beta_0 + \beta_1 x,$$

where $p(x)$ designates the probability $\mathbb{P}(Y = 1 | X = x)$.

Logistic regression belongs to the family of generalised linear models. An adjustment of these models in R is conducted using the **glm** function. Use of this function is similar to that of the **lm** function. A model must be written with a `formula` of the type (see Appendix A.2, p. 266):

$$Y \sim X_1 + X_2.$$

A family of probability distributions must also be chosen. Within the context of logistic regression, we are interested in binomial families:

```
> model_log.acid<-glm(Y~log.acid,data=prostate,family=binomial)
> summary(model_log.acid)
Call:
glm(formula = Y ~ log.acid, family = binomial, data = prostate)

Deviance Residuals:
    Min      1Q   Median      3Q     Max
-1.9802  -0.9095  -0.7265  1.1951  1.7302

Coefficients:
            Estimate Std. Error z value Pr(>|z|)
(Intercept)    0.404      0.509   0.794   0.4274
log.acid       2.245      1.040   2.159   0.0309 *
---
Signif. codes:  0  ***  0.001  **  0.01  *  0.05  .  0.1     1

(Dispersion parameter for binomial family taken to be 1)

    Null deviance: 70.252  on 52  degrees of freedom
Residual deviance: 64.813  on 51  degrees of freedom
AIC: 68.813

Number of Fisher Scoring iterations: 4
```

The output yields a `Coefficients` matrix with four columns for each parameter (row): estimations of parameters (column `Estimate`), estimated standard deviations of the parameters (`Std. Error`), observed values of the test statistic (`z value`), and p-values (`Pr(>|z|)`), which are the probability that, under the null hypothesis, the test statistics will exceed the estimated values of the parameters.

The coefficients β_0 and β_1 are estimated by the method of maximum likelihood. Here $\hat{\beta}_0 = 0.404$ and $\hat{\beta}_1 = 2.245$; we therefore obtain the model:

$$\log\left(\frac{\hat{p}(x)}{1 - \hat{p}(x)}\right) = 0.404 + 2.245x$$

or equivalently,

$$\hat{p}(x) = \frac{\exp(0.404 + 2.245x)}{1 + \exp(0.404 + 2.245x)}$$

It is possible to test the significance of the model's coefficients and to obtain confidence intervals for the same coefficients. For example, the p-value associated with the Wald test for hypotheses

$$H_0 : \beta_1 = 0 \qquad \text{against} \qquad H_1 : \beta_1 \neq 0$$

is 0.0309. The null hypothesis will therefore be rejected in favour of the alternative hypothesis with a 5% risk: the variable `log.acid` is retained in the model.

Figure 9.1 presents the estimated values of $p(x)$ obtained with the model `model_log.acid`:

```
> beta <- coef(model_log.acid)
> x <- seq(-2,2,by=0.01)
> y <- exp(beta[1]+beta[2]*x)/(1+exp(beta[1]+beta[2]*x))
> plot(x,y,type="l",xlab="log.acid",ylab="p(x)")
> abline(h=0.5,lty=2)
> xlim <- -beta[1]/beta[2]
> abline(v=xlim,lty=2)
```

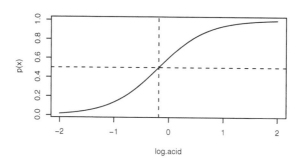

Figure 9.1
Estimated values of $p(x)$ for the model `model_log.acid`.

Note that

$$\hat{p}(x) \;=\; \widehat{\mathbb{P}}(Y = 1 | X = x) \begin{cases} \leq 0.5 & \text{if } x \leq -0.18 \\ > 0.5 & \text{if } x > -0.18 \end{cases}$$

For the values of x exceeding -0.18, the model will thus predict the value $\hat{Y} = 1$ (cancer has reached the lymphatic system) whereas for values less than or equal to -0.18, it will predict $\hat{Y} = 0$ (the cancer has not yet reached the lymphatic system).

It is of course possible to consider other logistic models. For example, the model with only the variable `size` as explanatory variable:

```
> model_size <- glm(Y~size,data=prostate,family=binomial)
```

```
> model_size
```

```
Call:  glm(formula = Y ~ size, family = binomial,
                                    data = prostate)
```

```
Coefficients:
(Intercept)          size1
    -1.435          1.658
```

```
Degrees of Freedom: 52 Total (i.e. Null);  51 Residual
Null Deviance:      70.25
Residual Deviance: 62.55  AIC: 66.55
```

In this case, only the coefficient associated with the level 1 of the variable `size` is estimated (the coefficient associated with the level 0 is by default taken to be zero). We obtain

$$\log\left(\frac{\hat{p}(x)}{1-\hat{p}(x)}\right) = \begin{cases} -1.435 & \text{if } x = 0 \\ -1.435 + 1.658 = 0.223 & \text{if } x = 1. \end{cases}$$

It can also be interesting to consider models comprising one or more interactions, such as that made up of the variables `age`, `grade` as well as the interaction `age:grade`:

```
> model_inter <- glm(Y~age+grade+age:grade,data=prostate,
      family=binomial)
```

The model which includes all the explanatory variables can also be studied:

```
> model <- glm(Y~.,data=prostate,family=binomial)
> model
```

```
Call:  glm(formula = Y ~ ., family = binomial, data = prostate)
```

```
Coefficients:
(Intercept)          age          acid         Xray1         size1
  10.08672      -0.04289      -8.48006       2.06673       1.38415
    grade1      log.acid
   0.85376       9.60912
```

```
Degrees of Freedom: 52 Total (i.e. Null);  46 Residual
Null Deviance:      70.25
Residual Deviance: 44.77  AIC: 58.77
```

3. Choosing the model:

Several models have already been estimated. Other models would of course also be possible. The **anova** function can be used to compare two nested models using the deviance statistic to test the nullity of the coefficients of the bigger model. For example, it is possible to compare the models `model` and `model_log.acid` using the command

```
> anova(model_log.acid,model,test="Chisq")
Analysis of Deviance Table

Model 1: Y ~ log.acid
Model 2: Y ~ age + acid + Xray + size + grade + log.acid
  Resid. Df Resid. Dev Df Deviance P(>|Chi|)
1        51     64.813
2        46     44.768  5   20.044  0.001226 **
---
Signif. codes:  0  ***  0.001  **  0.01  *  0.05  .  0.1  1
```

As the p-value of the test is 0.001226, the null hypothesis is rejected. Thus the model `model_log.acid` is rejected and at least one of the supplementary variables of `model` is considered to be relevant.

The **anova** function was used here to compare two models. Besides, the **step** function is used to choose a model with the help of a step-by-step procedure based on minimising the **AIC** value (Akaike Information Criterion):

```
> model_step <- step(model,direction="backward")
> model_step

Call:  glm(formula = Y ~ acid + Xray + size + log.acid,
                 family = binomial,data = prostate)

Coefficients:
(Intercept)         acid        Xray1        size1     log.acid
      9.067       -9.862        2.093        1.591       10.410

Degrees of Freedom: 52 Total (i.e. Null);  48 Residual
Null Deviance:      70.25
Residual Deviance: 46.43  AIC: 56.43
```

The step-by-step selection procedure conducted by the **step** function is here a backward procedure: at each step the variable for which withdrawal from the model leads to the greatest decrease of the **AIC** value is removed. The direction of the step-by-step procedure is specified by the argument `direction` (here `direction="backward"`). The process stops when all the variables are removed or when removing a variable no longer decreases the criterion. In this example, the variables age and grade have been removed. In order to use this backward method, it must first be ensured that the model used in the argument for the **step** function contains a sufficient number of variables. We could have used a more general model involving certain interactions as the initial model for the procedure. The **step** function can also be used to conduct forward or progressive (simultaneously forward and backward) step-by-step procedures.

4. Making a prediction:

The logistic models constructed previously can be used for prediction. The table below contains measurements from four new patients for the six explanatory variables:

Age	Acid	Xray	Size	Grade	log.acid
61	0.60	1	0	1	−0.51
49	0.86	0	0	1	−0.15
67	0.72	1	0	1	−0.33
51	0.95	1	1	1	−0.05

To obtain the values $p(x) = \mathbb{P}(Y = 1|X = x)$ predicted by the model model_step for these four new individuals, we first bring together the new data in a data-frame with the same structure as the initial data table (and especially the same names for the variables):

```
> new_d <- data.frame(matrix(c(61,0.60,1,0,1,-0.51,49,0.86,0,0,1,
 -0.15,67,0.72,1,0,1,-0.33,51,0.95,1,1,1,-0.05),ncol=6,byrow=T))
> names(new_d) <- names(prostate)[-6]
```

Remember, here the variables Xray, size, and grade must be converted into factors:

```
> for (i in 3:5){new_d[,i] <- factor(new_d[,i])}
```

The **predict** function is used to predict the probabilities $\mathbb{P}(Y = 1|X = x)$ for each of these new individuals:

```
> prediction <- predict(model_step,newdata=new_d,type="response")
> prediction
          1         2         3         4
0.4835626 0.2737392 0.6512557 0.9459345
```

For the first two individuals, the predicted probabilities are less than 0.5. We therefore predict $\hat{Y} = 0$, that is to say, the cancer has not reached the lymphatic system, whereas for the last two individuals we predict that the cancer has indeed reached the lymphatic system.

The **predict** function can also be used to estimate the model's misclassification rate (proportion of errors made by the model when applied to the available sample):

(a) First calculate the probabilities predicted for each individual from the sample:

```
> prediction_prob <- predict(model_step,newdata=prostate,
    type="response")
```

(b) Compare these probabilities to 0.5 in order to obtain the predicted label for each individual:

```
> prediction_label <- as.numeric(prediction_prob>0.5)
```

(c) Draw up the contingency table with the predicted labels and the true labels:

```
> table(prostate$Y,prediction_label)
   prediction_label
     0  1
  0 28  5
  1  6 14
```

We thus obtain the model's misclassification rate:

```
> error <- sum(prediction_label!=prostate$Y)/nrow(prostate)
> error
[1] 0.2075472
```

The rate obtained in this way is generally optimistic as the same sample is used to construct the model and to estimate the misclassification rate. We can obtain a more precise estimation using cross-validation methods. In order to do so, we use the function **cv.glm**. First of all, the package boot must be loaded and a cost function created which admits the observed values of Y as well as the predicted probabilities as input:

```
> library(boot)
> cost <- function(Y_obs,prediction_prob)
+    return(mean(abs(Y_obs-prediction_prob)>0.5))
> cv.glm(prostate,model_step,cost)$delta[1]
       1
0.2830189
```

9.2.5   Rcmdr Corner

1. Reading the data from a file:
Data → Import data → from text file, clipboard, or URL. The column separator (field separator) must then be specified.

Remember, here the variables `Xray`, `size`, `grade` and `Y` must be converted into factors: `Data → Manage variables in active data set → Convert numeric variables to factors....`
To check that the dataset has been imported successfully:
`Statistics → Summaries → Active data set`

2. Constructing the model:
`Statistics → Fit models → Generalized linear model....`
To write the model:

(a) Double-click on the response variable.

(b) Next choose the explanatory variables, separating them with a "+" and a ":" when an interaction is required.

3. Choosing the model:
`Models → Stepwise model selection`

The menu `Models` can also be used to obtain confidence intervals for the parameters of the model, to conduct tests between nested models, or even to represent residuals.

9.2.6   Taking Things Further

The **residuals** function can be used to calculate the different types of residuals coming from a logistic regression (Pearson or deviance residuals, for example). We might also use the function **logLik** to obtain the log-likelihood of a model or to calculate the value of the statistic for a likelihood ratio test. Should there be a response variable with more than two categories, two cases must be distinguished. When the categories are ordinal, we build an ordinal polytomous model using the **polr** function from the package MASS. When there is no ordinal relation between the categories of Y, we turn to a nominal polytomous model, also known as a multinomial model. In such cases we must use the **multinom** function from the package nnet. We can also call upon the **vglm** function of the package VGAM which can be used to build both ordinal polytomous models and multinomial models.

Furthermore, in order to make the prediction, we compared the probability $p(x)$ to the 0.5 threshold. It can often be wise to change the value of this threshold. In the example which we considered, we could choose a threshold lower than 0.5 if we wanted to minimise the risk of mistakenly predicting that cancer has not reached the lymphatic system. The techniques for scoring and tracing the set graphs such as ROC and lift curves (see package ROCR) can assist us in our choice of threshold. For further theoretical elements on logistic regression, refer to Fox and Weisberg (2011), Hosmer and Lemeshow (2000) and Collett (2003).

9.3 Decision Tree

9.3.1 Objective

Decision trees are tools for data exploration and to assist in decision-making, which can both explain and predict a quantitative variable (regression tree) or a qualitative variable (classification tree) from quantitative and/or qualitative explanatory variables. They consist of a sequential algorithm based on classification and regression tree (CART, Breiman et al., 1984), which constructs individual classes. The classes are generated using binary rules constructed from explanatory variables so that the individuals from a given class might be as homogeneous as possible in terms of the response variable.

The main advantages of this method are the simplicity of and ease with which the results can be both understood and interpreted due to the tree diagram produced. These trees are composed of "nodes" and "leaves" and are easy to read. Understanding is thus facilitated by labelling the nodes, which is made possible by producing decision rules. They are also used to select the relevant variables in a dataset with a great number of potentially interesting variables.

Two packages, rpart and tree, can be used to build CART-based trees. Here we will simply use the default package rpart.

9.3.2 Example

As an illustration we shall study the example of treating prostate cancer which is explained in detail in Worked Example 9.2. The objective remains the same: to predict Y (has the cancer reached the lymphatic system?) using the variables age, acid, Xray, size, grade and log.acid. The response variable is qualitative and we will therefore construct a classification tree.

9.3.3 Steps

1. Read the data.

2. Construct and analyse the decision tree.

3. Choose the size of the tree.

4. Estimate the classification error rate.

5. Predict Y for new individuals.

9.3.4 Processing the Example

1. Reading the data:

Import the data from file `prostate.txt` and convert the variables `Xray`, `size`, `grade` and `Y` into factors:

```
> prostate <- read.table("prostate.txt",header=T)
> for (i in 3:6) prostate[,i] <- factor(prostate[,i])
```

Summarise the dataset:

```
> summary(prostate)
      age              acid          Xray    size    grade   Y
 Min.   :45.00   Min.   :0.4000   0:38    0:26    0:33    0:33
 1st Qu.:56.00   1st Qu.:0.5000   1:15    1:27    1:20    1:20
 Median :60.00   Median :0.6500
 Mean   :59.38   Mean   :0.6942
 3rd Qu.:65.00   3rd Qu.:0.7800
 Max.   :68.00   Max.   :1.8700
    log.acid
 Min.   :-0.9163
 1st Qu.:-0.6931
 Median :-0.4308
 Mean   :-0.4189
 3rd Qu.:-0.2485
 Max.   : 0.6259
```

2. Constructing and analysing the decision tree:

We use the **rpart** function of the rpart package, which uses a formula like the **lm** function (see Appendix A.2, p. 266):

```
> library(rpart)
> prostate.tree <- rpart(Y~.,data=prostate)
> prostate.tree
n= 53

node), split, n, loss, yval, (yprob)
      * denotes terminal node

1) root 53 20 0 (0.6226415 0.3773585)
  2) acid< 0.665 28   4 0 (0.8571429 0.1428571) *
  3) acid>=0.665 25   9 1 (0.3600000 0.6400000)
    6) Xray=0 16   7 0 (0.5625000 0.4375000) *
    7) Xray=1 9   0 1 (0.0000000 1.0000000) *
```

Each node of the tree corresponds to a group of individuals, for which R yields

- Node number: `node`, for example, 3)

- Cut rule: `split`, for example, `acid>=0.665`
- Number of individuals in the group: `n`, for example, `25`
- Number of misclassified individuals: `loss`, for example, `9`
- Predicted value: `yval`, for example, `1`
- Probability of belonging to each class: `(yprob)`, for example `(0.3600 0.6400)`.

Thus in the case of the first node (called the "root" of the tree), it can be seen that there are 53 observations (all the individuals), and that for 62.3% of those individuals, cancer did not spread to the lymphatic system ($Y = 0$). This first node is segmented using the variable `acid`. The individuals are divided into two sub-groups: the first (node 2, `acid < 0.665`) is composed of 28 individuals, 85.7% of which correspond to $Y = 0$ and 14.3% to $Y = 1$. The second (node 3, `acid >= 0.665`) is composed of 25 individuals (36% $Y = 0$, 64% $Y = 1$). Node 2 is not segmented whereas node 3 is segmented according to the variable `Xray`, which yields nodes 6 and 7.

When the algorithm can no longer be segmented into nodes, we end by a leaf. Thus, in our example we obtain three leaves which correspond to nodes 2, 6 and 7, and which can be identified in the script by the symbol *. For a "pure" leaf, we obtain a group of individuals from just one class (example: in leaf 7, 100% of the individuals have a value of $Y = 1$).

Once the tree has been constructed, the class assigned to each leaf must of course be defined. For a pure leaf, the answer is obvious. If the leaf is not pure, a simple rule is to decide to assign it the class which is the most represented in the leaf. Thus, if we examine leaf 6 (56.25% $Y = 0$, 43.75% $Y = 1$), we can assume the following attribution rule: "if `acid > 0.665` and `Xray = 0` then $Y = 0$".

The **plot** function is used to obtain the tree which represents this segmentation graphically (Figure 9.2). A number of graphical parameters are used to obtain a tree which is easier to read:

```
> plot(prostate.tree,branch=0.2,compress=T,margin=0.1,
    main="Prostate.tree")
> text(prostate.tree,fancy=T,use.n=T,pretty=0,all=T)
```

This diagram summarises all the abovementioned information. There are 53 data items (33 `Y=0` and 20 `Y=1`) and therefore the predicted value at the first node is 0. We then read the segmentation rules (`acid < 0.665`, etc.) and the distribution of `Y=0` and `Y=1` in each node and each leaf. We can also calculate the number of misclassified data: here, $4 + 7 + 0 = 11$.

Prostate.tree

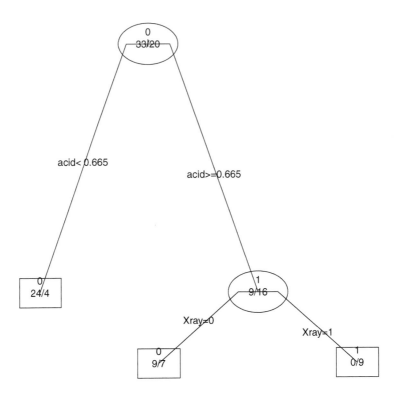

Figure 9.2
Decision tree: predict variable Y (prostate dataset).

To analyse a tree, it is possible to read all the results using the **summary** function. The advantage of this summary is that it lists the variables that compete with those that are chosen. For each node, the summary therefore uses `Primary splits` to return the primary variables which could have been chosen (as second, third or fourth choice depending on the `maxcompete` argument).

Here is an extract of the summary:

```
> prostate.tree <- rpart(Y~.,data=prostate,maxcompete=2)
> summary(prostate.tree)

Call:
rpart(formula = Y ~ ., data = prostate, maxcompete = 2)
  n= 53
```

```
     CP nsplit rel error xerror      xstd
1 0.35      0      1.00   1.00 0.1764428
2 0.10      1      0.65   0.90 0.1723861
3 0.01      2      0.55   0.85 0.1699056

Node number 1: 53 observations,      complexity param=0.35
  predicted class=0  expected loss=0.3773585
     class counts:     33     20
   probabilities: 0.623 0.377
  left son=2 (28 obs) right son=3 (25 obs)
  Primary splits:
     acid      < 0.665      to the left,  improve=6.528518, (0 missing)
     log.acid < -0.407996 to the left,  improve=6.528518, (0 missing)
     Xray      splits as  LR, improve=5.302152, (0 missing)
  Surrogate splits:
     log.acid < -0.407996 to the left,  agree=1.00, adj=1.00, (0 split)
     Xray      splits as  LR, agree=0.585, adj=0.12, (0 split)
     size      splits as  LR, agree=0.585, adj=0.12, (0 split)
     age       < 47.5      to the right, agree=0.566, adj=0.08, (0 split)

Node number 2: 28 observations
  predicted class=0  expected loss=0.1428571
     class counts:     24      4
   probabilities: 0.857 0.143
...
```

At the first node level, the two competing variables are log.acid and Xray, and the substitute variable is log.acid. It is clear that for the trees, the choice of the level at which the cuts of continuous variables are made is decided according to rank, which explains why the variables acid and log.acid provide exactly the same information. The surrogate splits can be used to handle missing values.

3. Choosing the size of the tree:

As the aim is to construct homogeneous groups, it may seem natural to choose the tree with the greatest number of pure leaves. However, with a view to predicting the Y value of a new individual, the challenge is to construct the tree which best represents reality. However, the more leaves a tree has, the greater the risk of misclassifying new individuals. On the contrary, the smaller the tree, the more stable its future predictions will be. There is thus a compromise to construct a tree that well represents the reality and is sufficiently stable. The **rpart** function always chooses trees through cross-validation using the argument xval: by default, this argument is fixed at 10 (the function constructs 10 samples of size $n/10$). Other arguments, such as the minimum number of data items found in a node, may interfere with this one (argument minsplit).

```
> prostate.tree2 <- rpart(Y~.,data=prostate,minsplit=5)
> plot(prostate.tree2,branch=0.2,compress=T,margin=0.1,
      main="Prostate.tree2")
> text(prostate.tree2,pretty=0)
```

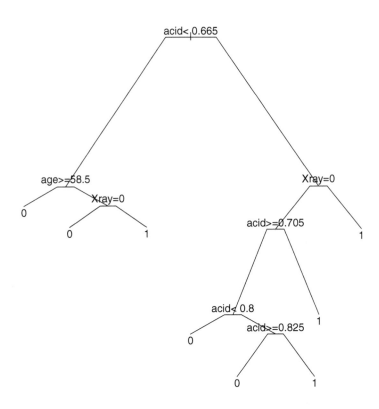

Prostate.tree2

Figure 9.3
Tree built using `minsplit=5`.

By modifying the `minsplit` argument, we obtain a bigger tree (Figure 9.3)
which must then be pruned.

The tree is pruned using the **snip.rpart** function, either directly with the mouse
in the graph window (which must be active) or by specifying the nodes to be
removed. Another method is to analyse the results of the cross-validation
using the **plotcp** function. We therefore conduct a classical cross-validation:
each individual is put aside and the tree is constructed with the remaining

$n - 1$ individuals and the initial individual is then predicted (the "leave-one-out" method).

```
> prostate.tree2 <- rpart(Y~.,data=prostate,minsplit=5,xval=53)
> plotcp(prostate.tree2)
```

We therefore choose a tree with a complexity value of `cp=0.087` and which also corresponds to a tree with four leaves (Figure 9.4).

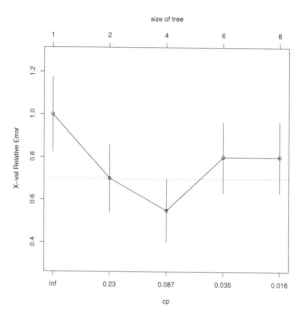

Figure 9.4
Choosing the size of the tree using **plotcp**.

Let us now construct the final tree:

```
> prostate.fin <- rpart(Y~.,data=prostate,minsplit=5,cp=0.087)
> plot(prostate.fin,branch=0.2,compress=T,margin=0.1,
      main="Prostate.fin")
> text(prostate.fin,fancy=T,use.n=T,pretty=0,all=T)
```

It must be noted here that only the variables `acid` and `Xray` are involved in predicting Y (Figure 9.5). Misclassifications are now equal to $4+3+0+0 = 7$.

4. Estimating the classification error rate:

To estimate the rate of misclassification of the final tree, we can read the number of classification errors on the graph in Figure 9.5, or calculate it directly using the object `prostate.fin`:

Prostate.fin

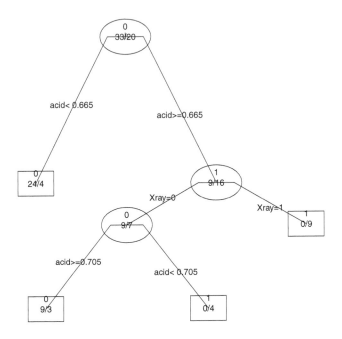

Figure 9.5
Final tree obtained using cross-validation.

```
> pred <- predict(prostate.fin,newdata=prostate,type="class")
> tab <- table(prostate$Y,pred)
> tab
   pred
     0  1
  0 33  0
  1  7 13
> error <- sum(pred!=prostate$Y)/nrow(prostate)
> error
[1] 0.1320755
```

Here the error rate of around 13.2% is calculated from the data which was used to construct the tree and therefore underestimates the real error rate even if the tree was chosen by cross-validation. To estimate the true rate, a forecast is made using cross-validation (the aforementioned leave-one-out method):

```
> pred <- prostate$Y
> for (i in 1:length(pred)){
+    tree <- rpart(Y~.,data=prostate[-i,],minsplit=1,cp=0.087)
```

```
+    pred[i] <- predict(tree,newdata=prostate[i,],type="class")
+ }
> tab <- table(prostate$Y,pred)
> tab
   pred
     0  1
  0 31   2
  1  9 11
> error <- sum(pred!=prostate$Y)/nrow(prostate)
> error
[1] 0.2075472
```

The estimated error rate is 20.8%. It can be seen that the rate is indeed greater than the misclassification rate estimated without cross-validation (around 13.2%).

5. Predicting Y for new individuals:

The above decision tree can be used in the context of forecasting. The following table contains measurements from four new patients for the six explanatory variables:

Age	Acid	Xray	Size	Grade	log.acid
61	0.60	1	0	1	−0.51
49	0.86	0	0	1	−0.15
67	0.72	1	0	1	−0.33
51	0.95	1	1	1	−0.05

To obtain the values of Y predicted by the decision tree for these four new individuals, we first bring together the new data in a data-frame with the same structure as the initial data table (and especially the same names for the variables):

```
> new.d <- data.frame(matrix(c(61,0.60,1,0,1,-0.51,49,0.86,0,0,1,
   -0.15,67,0.72,1,0,1,-0.33,51,0.95,1,1,1,-0.05),ncol=6,byrow=T))
> names(new.d) <- names(prostate)[-6]
> for (i in 3:5){new.d[,i] <- factor(new.d[,i])}
> summary(new.d)
```

```
      age              acid         Xray  size  grade     log.acid
 Min.   :49.0    Min.   :0.6000    0:1   0:3   1:4    Min.   :-0.510
 1st Qu.:50.5    1st Qu.:0.6900    1:3   1:1          1st Qu.:-0.375
 Median :56.0    Median :0.7900                       Median :-0.240
 Mean   :57.0    Mean   :0.7825                        Mean   :-0.260
 3rd Qu.:62.5    3rd Qu.:0.8825                       3rd Qu.:-0.125
 Max.   :67.0    Max.   :0.9500                        Max.   :-0.050
```

```
> pred <- predict(prostate.fin,newdata=new.d,type="class")
> pred
1 2 3 4
0 0 1 1
Levels: 0 1
```

Remark

Here, we must list all the explanatory variables for the new individuals. Another solution is to reconstruct the tree (with the **rpart** function) using only those variables which have previously been retained, prior to making the prediction.

For the first two individuals, the tree predicts $\hat{Y} = 0$, which means that the cancer has not yet reached the lymphatic system, whereas for the last two individuals, the tree predicts $\hat{Y} = 1$, that is, the cancer has reached the lymphatic system.

More precisely, for these individuals, we can also examine the probability of predicting $Y = 1$ (rather than $\hat{Y}$ directly):

```
> pred <- predict(prostate.fin,newdata=new.d)
> pred
            0          1
1 0.8571429 0.1428571
2 0.7500000 0.2500000
3 0.0000000 1.0000000
4 0.0000000 1.0000000
```

We can consider that the risk of being mistaken when we predict $\hat{Y} = 0$ for the first two individuals is rather low as, according to the tree, there is less than a 25% risk that the cancer has reached the lymphatic system.

9.3.5   Rcmdr Corner

It is not possible to construct a decision tree with Rcmdr.

9.3.6   Taking Things Further

It is possible to construct a tree when the variable Y is multivariate, again using **rpart**.

It is well known that decision trees are unstable. For this reason, we recommend the use of the argument **maxcompete** which displays the concurrent variables. It is also possible to combine a great number of trees using the randomForest package.

For more details on decision tree, refer to Breiman et al. (1984), Devroye et al. (1996) or Hastie et al. (2009).

10

Exploratory Multivariate Analysis

10.1 Principal Component Analysis

10.1.1 Objective

Principal component analysis (PCA) summarises a data table where the individuals are described by (continuous) quantitative variables. PCA is used to study the similarities between individuals from the point of view of all the variables and identifies individuals' profiles. It is also used to study the linear relationships between variables (from correlation coefficients). The two studies are linked, which means that the individuals or groups can be characterised by the variables and the links between variables illustrated from typical individuals.

We use the **PCA** function from the FactoMineR package (see Appendix A.4, p. 269) because it allows one to obtain numerous indicators and graphical outputs. Using this function, it is possible to add supplementary elements and construct the graphs simply and automatically.

10.1.2 Example

The dataset involves the results of decathlon events at two athletics competitions which took place one month apart: the Athens Olympic Games (23rd and 24th August 2004) and Decastar (25th and 26th September 2004). On the first day, the athletes compete in five events (100 m, long jump, shot put, high jump, 400 m) and in the remaining events on the second day (110 m hurdles, discus, pole vault, javelin, 1500 m). In Table 10.1, we have grouped together the performances for each athlete at each of the ten events, their overall rank, their final number of points and the competition in which they participated.

The aim of conducting PCA on this dataset is to determine profiles for similar performances: are there any athletes who are better at endurance events or those requiring short bursts of energy? And are some of the events similar? If an athlete performs well in one event, will he necessarily perform well in another?

TABLE 10.1

Athletes' Performances in ten Decathlon Events

	100m	Long	Shot	High	400m	110m	Disc	Pole	Jave	1500m	Rank	Points	Comp
Sebrle	10.85	7.84	16.36	2.12	48.36	14.05	48.72	5.00	70.52	280.01	1	8893	OG
Clay	10.44	7.96	15.23	2.06	49.19	14.13	50.11	4.90	69.71	282	2	8820	OG
Karpov	10.5	7.81	15.93	2.09	46.81	13.97	51.65	4.60	55.54	278.11	3	8725	OG
Macey	10.89	7.47	15.73	2.15	48.97	14.56	48.34	4.40	58.46	265.42	4	8414	OG
Warners	10.62	7.74	14.48	1.97	47.97	14.01	43.73	4.90	55.39	278.05	5	8343	OG
Zsivoczky	10.91	7.14	15.31	2.12	49.40	14.95	45.62	4.70	63.45	269.54	6	8287	OG
Hernu	10.97	7.19	14.65	2.03	48.73	14.25	44.72	4.80	57.76	264.35	7	8237	OG
⋮	⋮	⋮	⋮	⋮	⋮	⋮	⋮	⋮	⋮	⋮	⋮	⋮	⋮
SEBRLE	11.04	7.58	14.83	2.07	49.81	14.69	43.75	5.02	63.19	291.70	1	8217	Deca
CLAY	10.76	7.40	14.26	1.86	49.37	14.05	50.72	4.92	60.15	301.50	2	8122	Deca
⋮	⋮	⋮	⋮	⋮	⋮	⋮	⋮	⋮	⋮	⋮	⋮	⋮	⋮

10.1.3   Steps

1. Read the data.

2. Choose the active individuals and variables.

3. Choose if the variables need to be standardised.

4. Choose the number of dimensions.

5. Analyse the results.

6. Automatically describe the dimensions of variability.

7. Back to raw data.

10.1.4   Processing the Example

1. Reading the data:

It is here important to specify that the first column corresponds to the names of the individuals (`row.names=1`):

```
> decath <- read.table("decathlon.csv",sep=";",dec=".",
    header=TRUE, row.names=1)
> summary(decath)
```

2. Choosing the active individuals and variables:

As we want to determine the performance profiles, the variables corresponding to the athletes' performance are considered as active (the first ten variables).

The choice of active variables is very important: these variables, and these variables alone, will be taken into account when constructing the PCA dimensions. In other words, these are the only variables which are used to calculate the distances between individuals. As supplementary variables, we can add the quantitative variables `number of points` and `rank`, and the qualitative variable `competition`. Supplementary variables are very useful in helping to interpret the dimensions.

We can also choose the individuals which participate in constructing the dimensions, known as active individuals. Here, as is often the case, all the individuals are considered active.

3. Choosing if the variables need to be standardised:

We can centre and standardise the variables or simply centre them. For the dataset processed in this example, we do not have the choice. Standardisation is essential as the variables are of different units. When the variables are of the same units, both solutions are possible and lead to two separate analyses. This decision is therefore crucial. Standardisation means that every variable has the same importance. Not standardising the data therefore means that each variable has a weight corresponding to its standard deviation. This choice is all the more important as the variables have different variances.

To conduct PCA using the FactoMineR package (see Appendix A.4, p. 269), the preinstalled package must be loaded (see Section 1.6). We then use the **PCA** function.

```
> library (FactoMineR)
> res.pca <- PCA(decath, quanti.sup=11:12, quali.sup=13)
```

Variables 11 and 12 are quantitative supplementary variables and variable 13 is a qualitative supplementary variable. By default, the variables are centred and standardised (this is known as standardised PCA). To avoid scaling the variables, use the argument `scale.unit=FALSE`. By default, results are given for the first five dimensions. To specify another number of dimensions, use the argument `ncp`. The list `res.pca` contains all the results (see `names(res.pca)`).

4. Choosing the number of dimensions:

There are many solutions available in order to determine the number of dimensions to be analysed in PCA. The most common is to represent them using a bar chart of the eigenvalues or percentage of variance associated with each dimension (Figure 10.1):

```
> barplot(res.pca$eig[,2],names=paste("Dim",1:nrow(res.pca$eig)))
```

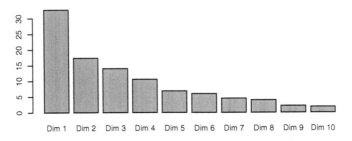

Figure 10.1
Percentage of variance associated with each dimension of the PCA.

We are therefore looking for a visible decrease or gap in the bar chart. Here, we can analyse the first four dimensions. Indeed, after four dimensions, we remark a regular decrease in the percentage of variance and can see a small "jump" between dimensions 4 and 5. To know the percentage of variance explained by the first four dimensions, we use the following line of code:

```
> round(res.pca$eig[1:4,],2)
        eigenvalue  percentage of  cumulative percentage
                        variance             of variance
comp 1      3.27          32.72                    32.72
comp 2      1.74          17.37                    50.09
comp 3      1.40          14.05                    64.14
comp 4      1.06          10.57                    74.71
```

The first two dimensions express 50% of the total percentage of variance, that is to say, 50% of the information in the data table is contained within the first two dimensions. This means that the diversity of the performance profiles cannot be summarised by two dimensions. Here, the first four dimensions are sufficient to explain 75% of the total variance.

5. Analysing the results:

By default, the PCA function yields the graph of individuals (Figure 10.2) and the graph of variables for the first two dimensions (Figure 10.3). The supplementary qualitative variable is represented on the graph of individuals. Each category of the qualitative variable competition is represented at the barycentre of the individuals which take this category. The supplementary quantitative variables are represented on the graph of variables. The correlation between these variables and the dimensions can be interpreted.

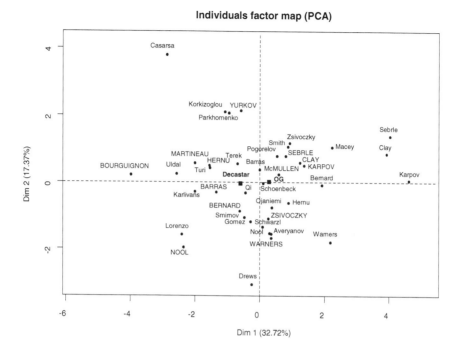

Figure 10.2
Decathlon PCA: graph of individuals (results of Décastar shown in capitals).

To interpret the results of a PCA, it is common to simultaneously study the results of the individuals and of the variables. Indeed, we want to characterise the different performance profiles from the events. The general rules for interpreting the two graphs simultaneously are the following ones: individuals which have high (resp. low) coordinates on a dimension take high (resp. low) values for the variables which are highly correlated to this dimension.

Overall, the first dimension confronts the "consistently high" performance profiles (i.e. of "good all-round" athletes), such as Karpov at the Olympics, and profiles of athletes who are "consistently weak" (relatively!), such as Bourguignon at Décastar.

It must be noted that in Figure 10.3, the variable 100m is negatively correlated with the variable long jump as athletes who run fast in the 100 metres, and therefore have a short time, jump a long way.

It must here be noted that the first principal component is the linear combination of the variables which best summarise all the variables. In this example, the automatic summary provided by PCA almost coincides with the number of points (the correlation is almost equal to 1), that is to say, the official

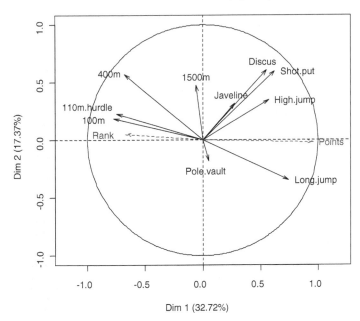

Figure 10.3
Decathlon PCA: graph of variables.

summary. Of course, the number of points was not used in the automatic summary, as this variable is purely illustrative.

We can interpret the position of the barycentres of the competition variable on the graph of individuals. The Olympics category has a positive coordinate on dimension 1 whereas the Décastar category has a negative coordinate. We can therefore consider that, on average, athletes perform better at the Olympics than at Décastar.

Dimension 2 is positively correlated to endurance variables (1500 m, 400 m) and power variables (discus, shot put). It means that dimension 2 opposes tough athletes (at the bottom since their time for 1500 m and 400 m is small) to powerful athletes on the top. It can be noted that both the categories of the qualitative variable `competition` have null coordinates on dimension 2, showing that overall there is no evolution between the Olympics and the Décastar from the second dimension point of view: this means that the athletes improved their overall performance (see dimension 1) but their profile did not change.

These interpretations can be refined using numerical results. The individuals' numerical results are contained within the object `res.pca$ind`. We then obtain a list with the individuals' coordinates on the dimensions (the principal components or scores), the squared cosines (which measure the quality of the projection of the individuals on the dimensions and are used to know if distances between individuals can be interpreted), and the contributions of the individuals to the construction of the dimensions. The results can be visualised by concatenating them; for example,

```
> round(cbind(res.pca$ind$coord[,1:3],res.pca$ind$cos2[,1:3],
    res.pca$ind$contrib[,1:3]), digits=2)
       Dim.1 Dim.2 Dim.3 Dim.1 Dim.2 Dim.3 Dim.1 Dim.2 Dim.3
Sebrle  4.04  1.37 -0.29  0.70  0.08  0.00 12.16  2.62  0.15
Clay    3.92  0.84  0.23  0.71  0.03  0.00 11.45  0.98  0.09
Karpov  4.62  0.04 -0.04  0.85  0.00  0.00 15.91  0.00  0.00
Macey   2.23  1.04 -1.86  0.42  0.09  0.29  3.72  1.52  6.03
   ⋮     ⋮     ⋮     ⋮     ⋮     ⋮     ⋮     ⋮     ⋮     ⋮
```

The list `res.pca$var` yields similar results for the variables: the coordinates, the squared cosines and the contributions of the variables. Loadings can be obtained by dividing the coordinates of the variables on a dimension by the square root eigenvalue of the corresponding dimension:

```
> loadings<-sweep(res$var$coord,2,sqrt(res$eig[1:5,1]),FUN="/")
```

Results of the supplementary quantitative variables are contained within the object `res.pca$quanti.sup` and the qualitative variables are contained within the object `res.pca$quali.sup`.

To improve the readability of the results, we can draw the graph of individuals by, for example, colouring the individuals according to a qualitative variable. In this case we use the **plot.PCA** function (which can be called indifferently using **plot** or **plot.PCA**, see Appendix A.1, p. 257). In the **plot.PCA** function, we provide the object which contains all the results of the PCA (`res.pca`), the graph that we want to construct (for the individuals `choix="ind"`) and we specify that the points are coloured according to the thirteenth variable (`habillage=13`). We can also alter the size of the font for the labels (`"cex = 0.7"` instead of 1 by default).

```
> plot(res.pca, choix="ind", habillage=13, cex=0.7)
```

We were only interested in the results of the first two dimensions, but it is possible to construct graphs of individuals and variables on dimensions 3 and 4:

```
> plot(res.pca, choix="ind", habillage=13, axes=3:4, cex=0.7)
> plot(res.pca, choix="var", habillage=13, axes=3:4)
```

To save a graph in pdf format, for example, we use the argument **new.plot=F**:

```
> pdf("mypath/mygraph.pdf")
> plot(res.pca, choix="var", habillage=13, axes=3:4, new.plot=F)
> dev.off()
```

6. Automatically describing the dimensions of variability:

In the FactoMineR package there is a function by which the dimensions of PCA can be described automatically: **dimdesc**. This function is a new aid to interpretation. It sorts the quantitative variables according to their coordinates on a dimension, that is to say, the correlation coefficient between the quantitative variable and the PCA dimension. Only the significant correlation coefficients are retained. This function also sorts the qualitative variables as well as the categories of the qualitative variables. In order to do so, a one-way analysis of variance model (see Section 8.1) is conducted for each qualitative variable: the response variable Y corresponds to the principal component (the coordinates of the individuals or scores) and is explained according to a qualitative variable. The p-value of the overall test (F-test) is calculated as well as the p-values of the tests of each category (test $H_0 : \alpha_i = 0$ with the constraint $\sum_i \alpha_i = 0$). Only the significant results are given. The p-values associated with the F-tests are sorted in ascending order. Thus, the qualitative variables are sorted from most to least characteristic. In the same way, the p-values of the t-tests by categories are sorted.

```
> dimdesc(res.pca)
$Dim.1
$Dim.1$quanti
                correlation       P-value
Points           0.9561543 2.099191e-22
Long.jump        0.7418997 2.849886e-08
Shot.put         0.6225026 1.388321e-05
High.jump        0.5719453 9.362285e-05
Discus           0.5524665 1.802220e-04
Rank            -0.6705104 1.616348e-06
X400m           -0.6796099 1.028175e-06
X110m.hurdle    -0.7462453 2.136962e-08
X100m           -0.7747198 2.778467e-09


$Dim.2
$Dim.2$quanti
                correlation       P-value
Discus           0.6063134 2.650745e-05
Shot.put         0.5983033 3.603567e-05
X400m            0.5694378 1.020941e-04
X1500m           0.4742238 1.734405e-03
```

```
High.jump    0.3502936 2.475025e-02
Javeline     0.3169891 4.344974e-02
Long.jump   -0.3454213 2.696969e-02
```

This function becomes very useful when there are a lot of variables. Here we can see that the first dimension is primarily linked to the variable Points (correlation coefficient of 0.96), and then to the variable 100m (negative correlation of −0.77), etc. The second dimension is principally described by the quantitative variables Discus and Shot put. None of the qualitative variables can be used to characterise dimensions 1 and 2 with a confidence level of 95%. The confidence level can be changed (proba = 0.2) and, for the qualitative variables, yields

```
> dimdesc(res.pca, proba = 0.2)
$Dim.1$quali
               P-value
Competition 0.1552515

$Dim.1$category
            Estimate    P-value
OG         0.4393744 0.1552515
Decastar  -0.4393744 0.1552515
```

7. Going back to raw data:

It is always important to refine the interpretation by going back to the raw data. As the raw data table is often difficult to read, to help us we can use the data table which has been centred-standardised:

```
> round(scale(decath[,1:12]),2)
          100m  Long  Shot  High  400m ...
Sebrle   -0.56  1.83  2.28  1.61 -1.09 ...
Clay     -2.12  2.21  0.91  0.94 -0.37 ...
Karpov   -1.89  1.74  1.76  1.27 -2.43 ...
Macey    -0.41  0.66  1.52  1.95 -0.56 ...
Warners  -1.44  1.52  0.00 -0.08 -1.43 ...
    ⋮      ⋮     ⋮     ⋮     ⋮     ⋮    ⋮
```

For example, Sebrle throws the shot put further than average (positive standardised value) somewhat remarkably (extreme value as it is greater than 2).

In the same way we can calculate the matrix of the correlations to know the exact correlation coefficient between the variables rather than the approximations provided by the graph:

```
> round(cor(decath[,1:12]),2)
         100m  Long  Shot  High  400m ...
100m     1.00 -0.60 -0.36 -0.25  0.52 ...
Long    -0.60  1.00  0.18  0.29 -0.60 ...
Shot    -0.36  0.18  1.00  0.49 -0.14 ...
High    -0.25  0.29  0.49  1.00 -0.19 ...
400m     0.52 -0.60 -0.14 -0.19  1.00 ...
          ⋮     ⋮     ⋮     ⋮     ⋮     ⋮  ⋮
```

10.1.5 Rcmdr Corner

A graphical interface for FactoMineR is also available in Rcmdr which makes it possible to conduct the previous analyses with an easy-to-use drop-down menu. To do this, simply install the drop-down menu once by launching the following command in an R window (internet connection required):

```
> source("http://factominer.free.fr/install-facto.r")
```

In following sessions, the FactoMineR drop-down menu will be present in Rcmdr and you will simply need to type **library(Rcmdr)**.

1. Import the data:

FactoMineR → Import data → from text file, clipboard, or URL ...

You must specify that the individuals' identifiers are available in the dataset. In order to do so, we select Row names in the first column. This option is available in the FactoMineR tab but is not available from the Data tab.

2–6. We will not discuss the choice of active elements and whether or not to standardise the variables, and will only show how to conduct PCA with the FactoMineR drop-down menu. Launch Rcmdr and click on the FactoMineR tab. Choose Principal Component Analysis to open the main window from the PCA drop-down menu (Figure 10.4).

By default, all the variables are considered as active, but it is possible to select some of them. It is also possible to select supplementary qualitative variables (Select supplementary factors), supplementary quantitative variables (Select supplementary variables) and supplementary individuals (Select supplementary individuals). By default, the results on the first five dimensions are provided in the object res, the variables are centred and standardised, and the graphs are provided for the first two dimensions. It is preferable to click Apply rather than Submit, as it launches the analysis whilst also keeping the window open, and it is thus possible to modify certain options without having to reset all the parameters.

The graphical options window (Figure 10.5) is separated into two parts. The left part concerns the graph of individuals and the right part the graph of variables. It is possible to represent only the supplementary qualitative variables

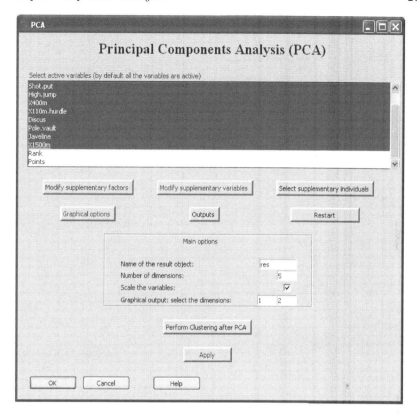

Figure 10.4
PCA main window in the FactoMineR menu.

(without the individuals `Hide some elements:` check ind); it is also possible to omit the labels for the individuals (`Label for the active individuals`). The individuals can be coloured according to a qualitative variable (`Coloring for individuals:` choose the qualitative variable).

The window for the different output options can be used to visualise the different results (eigenvalues, individuals, variables, automatic description of the dimensions). All the results can also be exported into a `csv` file (which can be opened in spreadsheet such as Excel, for example).

10.1.6 Taking Things Further

1. Constructing confidence ellipses around categories:

It is possible to construct confidence ellipses around the categories of a supplementary qualitative variable (see Figure 10.6). These ellipses thus give

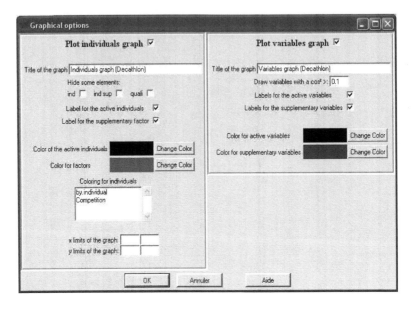

Figure 10.5
PCA graphical options window.

confidence regions for the position of each of the categories. To construct these ellipses, we write

```
> plotellipses(res.pca,keepvar=13)
```

2. Handling missing values in PCA:

It is possible to perform PCA with missing values using the missMDA package. The first step consists of imputing the missing values using the **imputePCA** function and the second step consists of performing PCA on the completed dataset:

```
> res <- imputePCA(my.incomplete.dataset)
> res.pca <- PCA(res$completeObs)
```

Overviews of PCA are available in many books such as Jolliffe (2002), Lebart et al. (1984) or Govaert (2009). In Husson et al. (2010), many studies conducted with FactoMineR are also described in detail.

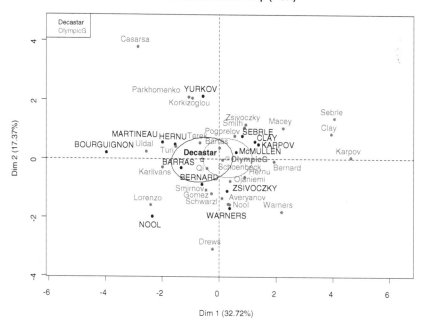

Figure 10.6
Confidence ellipses for the categories of a qualitative variable in PCA.

10.2 Correspondence Analysis

10.2.1 Objective

Correspondence analysis (CA) is used to summarise and visualise contingency tables, that is, a table confronting two qualitative variables. At the junction between row i and column j, we find the number of individuals carrying the category i of the first variable and j of the second. The aims of CA are as follows:

- Compare the row profiles with one another.

- Compare the column profiles with one another.

- Interpret the proximity between the rows and the columns, in other words, visualise the associations between the categories of the two variables.

To conduct a CA, we use the **CA** function from the FactoMineR package devoted to multivariate exploratory data analysis (see Appendix A.4, p. 269).

10.2.2 Example

The dataset represents the number of students from French universities by subject (or discipline) and by course by sex for the academic year 2007–2008. This is a table confronting the two qualitative variables Discipline and Level-sex. In the rows it features ten disciplines offered by the university and in the columns the junctions of the variables level (bachelors, masters and PhD) and sex (male or female). The CA is thus applied between one qualitative variable Discipline and a qualitative variable Level-sex corresponding to the confrontation of these two variables. This situation occurs frequently in correspondence analysis. Furthermore, for discipline, we also have the total number of students by level, by sex and an overall total (see Table 10.2).

The aim of this study is to have an overall image of the university. Which disciplines have the same student profiles? Which disciplines are favoured by women (and men respectively)? Which disciplines tend to have the longest courses of study?

10.2.3 Steps

1. Read the data.

2. Choose the active rows and columns.

3. Conduct the CA.

4. Choose the number of dimensions.

TABLE 10.2
University Datafile

	Bachelors		Masters		...	Sum
	F	M	F	M		
Law, Political Science	69373	37317	42371	21693	...	179125
Social Science, Economics, Management	38387	37157	29466	26929	...	136474
Economic and Social Administration	18574	12388	4183	2884	...	38029
Literature, Linguistics, Arts	48691	17850	17672	5853	...	96998
Languages	62736	21291	13186	3874	...	103833
Social Sciences and Humanities	94346	41050	43016	20447	...	213618
Joint Humanities	1779	726	2356	811	...	5700
Mathematics, Physics	22559	54861	17078	48293	...	158689
Life Science	24318	15004	11090	8457	...	69742
Sport Science	8248	17253	1963	4172	...	32152

5. Analyse the results.

10.2.4 Processing the Example

1. Reading the data:

It is here important to specify that the first column corresponds to the names of the rows (`row.names=1`). Check that the data has been imported correctly using the **summary** function or by visualising an extract of the dataset.

```
> University <- read.table("University.csv",sep=";",
    header=T, row.names=1)
> summary(University)
> University[1:5,1:3]
                            Bachelors.F Bachelors.M Masters.F
Law, Political Science            69373       37317     42371
Social Sc., Eco., Management      38387       37157     29466
Economic and Social Admin.        18574       12388      4183
Literature, Linguistics, Arts     48691       17850     17672
Languages                         62736       21291     13186
```

2. Choosing the active rows and columns:

The choice of active rows and columns is vital as only these rows and columns are taken into account when constructing the dimensions. As we want to compare all the disciplines, we will consider them as active rows. The columns for the variable `Level-sex` will be considered active and those corresponding to the various totals will be considered illustrative (the information contained within these variables is already taken into account by the active columns). These variables (from 7 to 12) are only used for interpreting the results.

3. Conducting the CA:

The correspondence analysis can now be conducted using the FactoMineR package (see Appendix A.4, p. 269). In order to do so, the package must be installed and then loaded (see Section 1.6). We then use the **CA** function.

```
> library(FactoMineR)
> res.ca <- CA(University, col.sup=7:12)
```

As mentioned above, columns 7 to 12 are considered illustrative. The list `res.ca` contains all the outputs. By default, a chi-square test is constructed to test the independence of the variables `Disciplines` and `Level-sex`. The test is constructed using only the active categories (rows and columns). In our example the independence hypothesis is largely rejected as the p-value is near 0: The chi-square of independence between the two variables is equal to 170789.2 (p-value = 0).

This means, as might be expected, that there are associations between certain disciplines and sex-level combinations. It is therefore interesting to construct a CA in order to visualise these associations between categories.

4. Choosing the number of dimensions:

In order to determine the number of dimensions to be analysed in CA, it can be useful to represent the eigenvalues or the percentage of variance associated with each dimension in bar charts (Figure 10.7).

```
> barplot(res.ca$eig[,2],names=paste("Dim",1:nrow(res.ca$eig)))
```

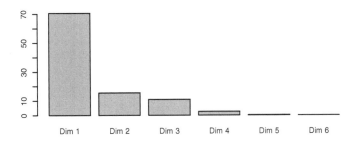

Figure 10.7
Percentage of variance associated with each dimension of the CA.

We can also directly examine the percentages of variance associated with each dimension using the following command:

```
> round(res.ca$eig,3)
      eigenvalue  percentage of  cumulative percentage
                     variance          of variance
```

dim 1	0.117	70.718	70.718
dim 2	0.026	15.508	86.226
dim 3	0.018	10.901	97.127
dim 4	0.004	2.627	99.754
dim 5	0.000	0.246	100.000
dim 6	0.000	0.000	100.000

The percentages of variance associated with the first dimensions are high, and we can simply describe the first three dimensions, as confirmed by the jump seen in Figure 10.7. The first three dimensions express nearly 97% of the total variance: In other words, 97% of the information in the data table is summarised by the first three dimensions.

5. Analysing the results:

By default, the **CA** function yields a graph simultaneously representing (active and illustrative) rows and columns on the first two dimensions. In Figure 10.8

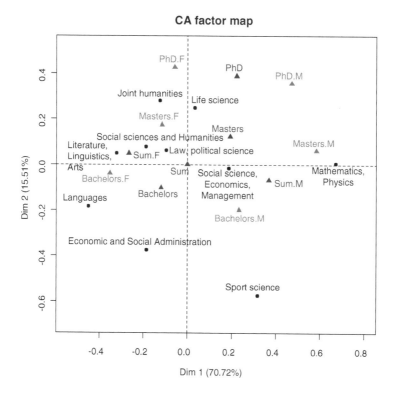

Figure 10.8
CA on university data.

we can view the major tendencies emanating from the analysis. If the graph contains too many points and becomes difficult to read, we can, for example, represent the rows alone using the invisible argument of the **plot.CA**:

```
> plot(res.ca, invisible=c("col","col.sup"))
```

First of all it must be noted that, by design, there can be no size effect in CA as each row (and column respectively) is divided by its margin. Thus, in the example, we *do not see* on one side the disciplines which attract a lot of students and on the other those which attract the fewest.

We can also examine proximities between the disciplines. Two disciplines are considered similar if they have the same profiles (they attract students of the same sex and for the same course length). For example, the graph shows that the disciplines Languages and Literature, Linguistics, Arts mainly attract women at Bachelors level. The supplementary columns can help in interpreting the CA graph. Thus, the first dimensions opposes men and women, whereas the second dimensions classes the levels from Bachelors (at the bottom) to PhD (at the top). The disciplines on the left of the graph (or right, respectively) are mainly studied by women (or men, respectively): literary disciplines are towards the left of the graph (and therefore studied mainly by women) and scientific disciplines are towards the right of the graph (and thus studied by men). The disciplines towards the bottom of the graph are those with short courses (Bachelor level over-represented by social and economic administration and sport, for example), whereas the disciplines towards the top of the graph have longer courses (PhD level over-represented for biological and life sciences, for example). However, it is not always easy to interpret the dimensions of the CA. In this case, we focus on the proximity between the categories.

To refine these interpretations, we can go on to examine all the numerical results. The numerical results of the rows are contained within the object res.ca$row. We thus obtain a table with the row coordinates on the dimensions, a table with the cosines squared (which represent the quality of the projection of a row category), and a table with the contributions of the rows to the construction of the dimensions (not given here):

```
> round(cbind(res.ca$row$coord[,1:3],res.ca$row$cos2[,1:3]),2)
```

	Dim 1	Dim 2	Dim 3	Dim 1	Dim 2	Dim 3
Law, political science	-0.10	0.07	-0.13	0.30	0.13	0.55
Social sc., Eco., Management	0.18	-0.02	-0.20	0.46	0.00	0.52
Economic and Social Admin.	-0.19	-0.37	0.01	0.20	0.80	0.00
Literature, Ling., Arts	-0.32	0.05	0.08	0.91	0.02	0.06
Languages	-0.45	-0.18	0.09	0.79	0.13	0.03
Social sc. and Humanities	-0.19	0.08	0.01	0.84	0.15	0.00
Joint humanities	-0.13	0.28	-0.58	0.04	0.18	0.77
Mathematics, Physics	0.67	0.01	0.08	0.98	0.00	0.01

| Life science | 0.03 | 0.25 | 0.28 | 0.01 | 0.41 | 0.52 |
| Sport science | 0.32 | -0.57 | 0.08 | 0.21 | 0.67 | 0.01 |

Similarly, the column results are obtained using

```
> round(cbind(res.ca$col$coord[,1:3],res.ca$col$cos2[,1:3],
    res.ca$col$contrib[,1:3]),2)
```

	Dim 1	Dim 2	Dim 3	Dim 1	Dim 2	Dim 3	Dim 1	Dim 2	Dim 3
Bachel.F	-0.35	-0.04	0.04	0.96	0.01	0.01	39.72	2.27	3.65
Bachel.M	0.23	-0.20	0.03	0.55	0.39	0.01	11.51	37.49	1.21
Masters.F	-0.11	0.17	-0.21	0.14	0.33	0.49	1.99	20.90	43.75
Masters.M	0.58	0.06	-0.07	0.95	0.01	0.01	40.43	1.97	3.26
PhD.F	-0.06	0.43	0.39	0.01	0.49	0.42	0.08	21.02	25.41
PhD.M	0.47	0.36	0.35	0.46	0.26	0.26	6.27	16.36	22.71

We can then examine the results concerning the supplementary elements contained within the object `res.ca$col.sup`:

```
> round(cbind(res.ca$col.sup$coord[,1:3],
    res.ca$col.sup$cos2[,1:3]),2)
```

	Dim 1	Dim 2	Dim 3	Dim 1	Dim 2	Dim 3
Sum.F	-0.26	0.05	-0.02	0.44	0.02	0.00
Sum.M	0.37	-0.07	0.02	0.60	0.02	0.00
Bachelors	-0.12	-0.10	0.04	0.13	0.09	0.01
Masters	0.19	0.12	-0.15	0.23	0.10	0.13
PhD	0.22	0.39	0.37	0.11	0.35	0.32
Sum	0.00	0.00	0.00	0.00	0.00	0.00

As expected, the coordinates of the total are at the barycentre of the cloud. The cloud's centre of gravity corresponds to the mean profile. This shows that if the position of a `Level-sex` combination is close to the centre of the cloud, this combination has the same discipline profile as that of the students overall.

We can also construct the graph for dimensions 3 and 4:

```
> plot(res.ca, axes = 3:4)
```

10.2.5   Rcmdr Corner

A graphical interface is also available in Rcmdr which makes it possible to conduct the previous analyses with an easy-to-use drop-down menu. To do this, simply install the drop-down menu once by launching the following command in an R window (Internet connection required):

```
> source("http://factominer.free.fr/install-facto.r")
```

In following sessions, the FactoMineR drop-down menu will be present in Rcmdr and you will simply need to type **library(Rcmdr)**.

1. Reading the data from a file:

FactoMineR → Import data → from text file, clipboard, or URL ...

You must specify that the row names are available in the dataset. In order to do so, we select Row names in the first column. This option is available in the FactoMineR tab but is not available from the Data tab.

2–5. We will not describe the choice of active elements but simply show how to conduct CA in Rcmdr.

To conduct CA, click on the FactoMineR tab and then on Correspondence analysis. In the main window of the CA drop-down menu, it is possible to select supplementary rows (Select supplementary rows), or columns (Select supplementary columns). By default, results on the first five dimensions are provided in the object res. It is preferable to click Apply (rather than Submit), which launches the analysis whilst keeping the window open, and it is thus possible to modify certain options without having to reconfigure all the parameters.

In graph options it is possible to omit the representation of certain elements (rows or columns, active or supplementary) or labels. The additional output options window is used to obtain the different results (eigenvalues, rows, columns). All the results can also be exported into a csv file (which can be opened in spreadsheet such as Excel, for example).

10.2.6   Taking Things Further

1. Relationship with the χ^2:

It can be interesting to go back to the raw data to analyse in more detail the relationship between two variables and particularly the proximity between the categories of these variables. We can also construct a chi-square test (see Worked Example 6.2) and calculate the matrix of deviations from independence.

```
> resultat <- chisq.test(University[,1:6])
> round(100 * resultat$residuals^2 / resultat$stat, 2)
> round(resultat$residuals,1)
```

This matrix of deviation from independence is visualised using CA.

2. Text mining:

Correspondence analysis is an extremely useful tool for conducting text analyses. Text analysis consists of comparing texts, for example, from two different authors, based on the words used in the texts. The principle is as follows: We list all the words used in the texts and then draw up a table with the texts in the rows and the words in the columns. Within a cell, we specify the number of times a word has been used in a text. Applying a CA to this kind of table makes it possible to compare the two authors and to see which words are over- or under-used by which texts.

Overviews of CA can be found in Greenacre and Blasius (2006), Greenacre (2007), Lebart et al. (1984), Murtagh (2005) or Govaert (2009) as well as in Husson et al. (2010). The last book also details different case studies conducted using FactoMineR.

10.3   Multiple Correspondence Analysis

10.3.1   Objective

The aim of multiple correspondence analysis (MCA) is to summarise and visualise a data table where the individuals are described by qualitative variables. MCA is used to study the similarities between individuals from the point of view of all the variables and identify individuals' profiles. It is also used to assess the relationships between the variables and study the associations of the categories. Finally, as with PCA and CA, the individuals or groups of individuals (rows) can be characterised by the categories of the variables (columns). If certain variables from the data table are quantitative, it is possible to include them in the analysis by dividing them into classes (Section 2.3.2, p. 35). To conduct MCA, we use the **MCA** function from the FactoMineR package specially designed for multivariate exploratory data analysis (see appendix A.4, p. 269).

10.3.2   Example

The dataset contains information from sixty-six clients who took out loans from a credit company. The eleven qualitative variables and the associated categories are as follows:

- `Loan`: Renovation, Car, Scooter, Motorbike, Furnishings, Sidecar. This variable represents the item for which clients took out a loan.

- `Deposit`: yes, no. This variable indicates whether or not clients paid a deposit before taking out the loan. A deposit represents a guarantee for the loan organisation.

- `Unpaid`: 0, 1 or 2, 3 and more. This variable indicates the number of unpaid loan repayments for each client.

- `Debt load`: 1 (low), 2, 3, 4 (high). This variable indicates the client's debt load. The debt load is calculated as the ratio between costs (sum of expenses) and income. The debt load is divided into four classes.

- `Insurance`: No insurance, DI (Disability Insurance), TPD (Total Permanent Disability), Senior (for people older than 60). This variable indicates the type of insurance the client has taken out.

- `Family`: Common-law, Married, Widower, Single, Divorcee.

- `Children`: 0, 1, 2, 3, 4 and more.

- `Accommodation`: Home owner, First-time buyers, Tenant, Lodged by family, Lodged by employer.

- **Profession**: Technician, Manual Labourer, Retired, Management, Senior Management.

- **Title**: Mr., Mrs., Miss.

- **Age**: 20 (18 to 29 years old), 30 (30 to 39), 40 (40 to 49), 50 (50 to 59), 60 and over.

The aim of this study is to characterise the credit company's clientele. Firstly we want to identify the different banking behaviour profiles, that is, to typify the individuals. We then study the relationship between the labels (profession, age, etc.) and the dimensions of variability for the banking behaviour profiles (i.e. characterise the clients with specific behaviours).

10.3.3   Steps

1. Read the data.

2. Choose the active individuals and variables.

3. Choose the number of dimensions.

4. Analyse the results.

5. Automatically describe the dimensions of variability.

6. Back to raw data in cross tabulation.

10.3.4   Processing the Example

1. Reading the data:

```
> Credit <- read.table("Credit.csv",sep=";", header=TRUE)
```

By examining the results yielded by the **summary** function, we remark that the Age variable is not considered qualitative. It therefore needs to be converted:

```
> Credit[,"Age"] <- factor(Credit[,"Age"])
```

In MCA it is important to check that there are no rare categories, that is, those with very small frequencies, as MCA attributes a lot of importance to these categories. To do this, we recommend representing each variable:

```
> for (i in 1:ncol(Credit)){
+    par(ask=TRUE)  # plot graphs one-by-one
+    plot(Credit[,i])
}
```

If certain variables admit categories with small frequencies, there are a number
of ways to deal with it:

- Naturally grouping together certain categories; we recommend this solution for ordinal categories (see Exercise 2.7).
- Ventilation (random allocation) of the individuals associated with rare
  categories in the other categories (see Exercise 2.6); certain programs
  offer automatic ventilation (see below).
- Elimination of individuals with rare categories.

In this example, one single individual carries the category `Sidecar` for the
`Loan` variable. It is therefore only natural to group this category along with
`Motorbike` (see Section 2.3.3, p. 37):

```
> levels(Credit[,"Loan"])[6] <- "Motorbike"
```

2. Choosing the active individuals and variables:

We want to determine banking behaviour profiles and will thus indicate as active those variables corresponding to banking information (the first five). The
choice of active variables is very important as only these variables contribute
to the construction of the MCA dimensions, and thus only these variables
are used to calculate the distances between individuals. As supplementary
variables, we add the other qualitative variables (corresponding to the questions about general information for the clients). These variables are useful
in interpreting the results. It is also possible to add quantitative variables as
supplementary (without having to divide them into classes in advance).

We can also choose the individuals who participate in constructing the dimensions, known as active individuals. Here, as is often the case, all the
individuals are considered active.

To conduct the MCA using the FactoMineR package (see Appendix A.4, p. 269),
the preinstalled package must be loaded (see Section 1.6). We then use the
MCA function.

```
> library(FactoMineR)
> res.mca <- MCA(Credit, quali.sup=6:11, level.ventil=0)
```

The MCA is constructed from the first five variables (the active variables)
whereas the variables 6 to 11 are supplementary. The argument `level.ventil`
is 0 by default, which means that no ventilation is conducted. If the argument
has a value of 0.05, this means that the categories with frequency equal to or
less than 5% of the total number of individuals are ventilated automatically
by the **MCA** function prior to constructing the MCA. By default, results are
given for the first five dimensions. To specify another number of dimensions,
use the argument `ncp`. The list `res.mca` contains all the results.

3. Choosing the number of dimensions:

There are many solutions available in order to determine the number of dimensions to be analysed in MCA. The most well-known is to represent the barplot of eigenvalues or of percentages of variance associated with each dimension.

> **barplot(res.mca$eig[,2],names=paste("Dim",1:nrow(res.mca$eig)))**

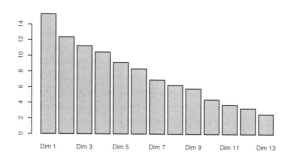

Figure 10.9
Percentage of variance associated with each dimension of the MCA.

The decrease in percentages of variance (Figure 10.9) is regular. There is no particularly noticeable gap or break in the diagram, and it is therefore difficult to choose the number of dimensions. However, these low levels of percentage of variance and the decrease are common in MCA. To obtain the numerical results, we use the following instruction:

> **round(res.mca$eig[1:5,],2)**

	eigenvalue	percentage of variance	cumulative percentage of variance
dim 1	0.40	15.33	15.33
dim 2	0.32	12.38	27.71
dim 3	0.29	11.25	38.96
dim 4	0.27	10.47	49.43
dim 5	0.24	9.14	58.57

The first two dimensions express 28% of the total variance. In other words, 28% of the information in the data table is summarised by the first two dimensions, which is relatively high in MCA. We will interpret only the first two dimensions even though it may be interesting to analyse dimensions 3 and 4.

4. Analysing the results:

The **MCA** function yields a graph simultaneously representing individuals and categories (active and illustrative) on the first two dimensions. This representation quickly becomes saturated and it is therefore necessary to conduct separate representations for the individuals and the categories of the variables. For this we use the `invisible` argument of the **plot.MCA** function. To represent the individuals only, we use

```
> plot(res.mca, invisible=c("var","quali.sup"))
```

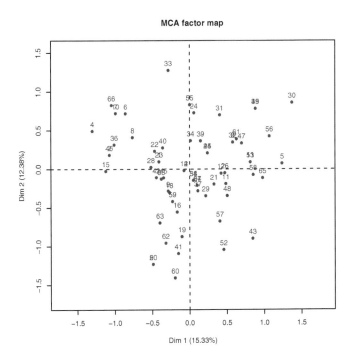

Figure 10.10
MCA of credit data: representation of the individuals.

This graph of individuals (Figure 10.10) illustrates the general shape of the scatterplot to see if, for example, it identifies particular groups of individuals. This is not the case here. To make the results easier to interpret, it can be helpful to colour the individuals according to a variable, with one colour used for each category of the variable. We thus use the `habillage` argument specifying either the name or the number of the qualitative variable (graph not provided):

```
> plot(res.mca, invisible=c("var","quali.sup"),habillage="Loan")
> plot(res.mca, invisible=c("var","quali.sup"),habillage=1)
```

The graph with all the categories of the active and illustrative qualitative variables (Figure 10.11) can be used to get an idea of the overall tendencies emanating from the analysis. To construct this graph and represent the categories alone, we use

```
> plot(res.mca, invisible="ind")
```

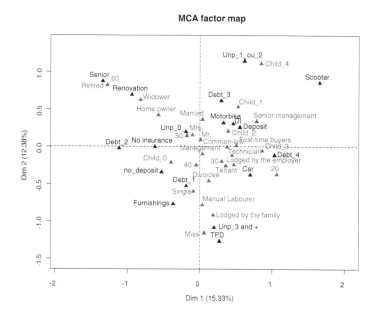

Figure 10.11
MCA of credit data: representation of active and illustrative categories.

The first dimension of Figure 10.11 opposes a "young" profile (on the right) with an "old" profile (on the left). We thus find older people, home-owners who have taken out a loan to finance renovation work, opposing younger people who have taken out a loan, for example, to buy a scooter. "Young" people tend to have a relatively high debt load.

Dimension 2 mainly opposes people in great financial difficulty (below) with the others (above). We can therefore identify people who have difficulty paying back their loan (three or more payments missed) and those who have taken out TPD insurance. This dimension can thus be qualified as the dimension of financial difficulty.

Using a number of examples, we evoke the general rules for interpreting the proximity between categories of two variables, be they the same or different. For example, the category Senior of the variable Insurance is close to the category `Retired` of the (illustrative) variable `Profession`, thus indicating that many of the individuals who carry the category `Senior` also carry the category `Retired`. We can also see that the category `Lodged by the employer` and `Tenant` for the `Accommodation` variable are close, which means that these people generally carry the same categories for the other variables. The individuals who carry these categories thus have the same profile.

To interpret a dimension in more detail, we recommend analysing the numerical data of the variables contained within the object `res.mca$var`. We thus obtain a list with the coordinates of the categories on the dimensions `res.mca$var$coord`, the squared cosines `res.mca$var$cos2` (which measure the quality of the projection of the categories on the dimensions) and the contributions of the categories to the construction of the dimensions `res.mca$var$contrib`. We also obtain `res.mca$var$eta2` which, for each dimension, corresponds to the correlation ratio η^2 between the principal component (the coordinates of the individuals on the dimension) and each qualitative variable. The individuals' numerical results are contained within the object `res.mca$ind`. The coordinates of the individuals (scores or principal components) are contained within the object `res.mca$ind$coord`.

To know which variables are most related to the dimensions, that is, to interpret the dimensions from the variables, we draw the graph of correlation ratios between the dimensions and the variables (Figure 10.12):

```
> plot(res.mca, choix="var")
```

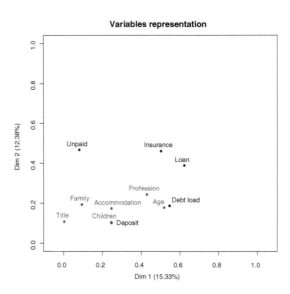

Figure 10.12
MCA of credit data: representation of active and illustrative variables.

The variable which contributes the most to the creation of dimension 1 is the Loan variable, and that which contributes the most to the creation of dimension 2 is the Unpaid variable (Figure 10.12). This information summarises the overall influence of all the categories of each of the variables to the construction of the dimensions. In this way, the categories of the Loan variable contribute greatly to the creation of dimension 1 and the categories of the Unpaid variable to the creation of dimension 2. We can then examine the results concerning the supplementary qualitative variables contained within the object res.mca$quali.sup.

It is possible to construct the graphs of individuals and variables on dimensions 3 and 4:

```
> plot(res.mca, invisible="ind", axes = 3:4)
> plot(res.mca, invisible=c("var","quali.sup"), axes=3:4)
> plot(res.mca, choix="var", axes=3:4)
```

To save a graph in pdf format for example, we use the argument new.plot=F:

```
> pdf("mypath/mygraph.pdf")
> plot(res.mca,invisible=c("var","quali.sup"),axes=3:4,
    new.plot=FALSE)
> dev.off()
```

5. Automatically describing the dimensions of variability:

In the FactoMineR package there is a function by which the dimensions of MCA can be described automatically: **dimdesc**. This function is a new aid to interpretation. It sorts the supplementary quantitative variables according to their coordinates on a dimension, that is to say, the correlation coefficient between the quantitative variable and the MCA dimension. Only the significant correlation coefficients are retained. This function also sorts the qualitative variables as well as the categories of the qualitative variables. In order to do so, a one-way analysis of variance model (see Section 8.1) is constructed for each qualitative variable: variable Y corresponds to the principal component (the coordinates of the individuals or scores) and is explained according to a qualitative variable. The p-value of the overall test (F-test) is calculated as well as the p-values of the tests of each category (test $H_0 : \alpha_i = 0$ with the constraint $\sum_i \alpha_i = 0$). Only the significant results are given. The p-values associated with the F-tests are sorted in ascending order. Thus, the qualitative variables are sorted from most to least characteristic. In the same way, the p-values of the t-tests by categories are sorted (in ascending order when the coefficient is positive, and in descending order when it is negative). This will identify the most characteristic categories. Below is a description of the second dimension of MCA from the variables and the categories:

```
> dimdesc(res.mca)
```

```
$`Dim 2`
$`Dim 2`$quali
                        R2        p.value
Unpaid          0.4679498 2.330816e-09
Insurance       0.4620452 1.979222e-08
Loan            0.3895658 3.732164e-06
Profession      0.2440563 1.661355e-03
Debt.load       0.1865282 4.857269e-03
Deposit         0.1033137 8.498296e-03
Family          0.1945394 9.447136e-03
Age             0.1781788 1.618904e-02
Accommodation   0.1734846 1.883183e-02
Title           0.1101503 2.532061e-02


$`Dim 2`$category
                        Estimate        p.value
Unp_1_ou_2             0.5997078 6.148708e-07
Senior                 0.5149935 2.212144e-05
Retired                0.4477685 1.257211e-03
60                     0.4556995 1.538945e-03
Debt_3                 0.3550533 1.640259e-03
Renovation             0.3127094 4.448861e-03
Home owner             0.3432251 6.371855e-03
Deposit                0.1826945 8.498296e-03
Scooter                0.4031618 1.373824e-02
DI                     0.1652700 4.212931e-02
Widower                0.3640249 4.252309e-02
Lodged by the family  -0.4156119 3.553903e-02
Child_0               -0.2990457 2.172157e-02
Single                -0.3322199 1.086364e-02
Debt_1                -0.2911376 1.009127e-02
no_deposit            -0.1826945 8.498296e-03
Miss                  -0.4824068 8.440543e-03
Car                   -0.2954665 8.029752e-03
Manual Labourer       -0.4605202 1.323161e-03
Furnishings           -0.5179085 1.049530e-05
TPD                   -0.6985157 3.360963e-09
Unp_3 and +           -0.6634535 7.988517e-10
```

This function becomes very useful when there are a lot of categories. Here we find that the second dimension is mainly linked to the categories TPD and Unp_3 and +.

6. Going back to raw data in cross-tabulation:

It is interesting to go back to the raw data to analyse more closely the relationship between two variables and particularly the proximity between the categories of these variables. We can thus construct cross-tabulations using the **table** command, calculate row and column percentages and conduct a chi-square test (see Worked Example 6.2).

10.3.5 Rcmdr Corner

A graphical interface for FactoMineR is also available in Rcmdr which makes it possible to conduct the previous analyses with an easy-to-use drop-down menu. To do this, simply install the drop-down menu once by launching the following command in an R window (Internet connection required):

```
> source("http://factominer.free.fr/install-facto.r")
```

In following sessions, the FactoMineR drop-down menu is present in Rcmdr and you simply need to type **library**(Rcmdr).

1. Import the data:

```
Data → Import data → from text file, clipboard, or URL ...
```

2–6. We will not describe the choice of active elements but simply show how to conduct MCA in Rcmdr.

Click on the FactoMineR tab and then Multiple Correspondence analysis. In the main MCA window, it is possible to select supplementary qualitative variables (Select supplementary factors), supplementary quantitative variables (Select supplementary variables) and supplementary individuals (Select supplementary individuals). By default, results on the first five dimensions are provided in the object res. It is preferable to click Apply rather than Submit, as it launches the analysis whilst also keeping the window open, and it is thus possible to modify certain options without having to reset all the parameters.

The graph options window and the other output options window are similar to those for PCA. In graph options it is possible to omit the representation of certain elements (individuals or categories) or labels.

The window for the output options can be used to visualise the different results (eigenvalues, individuals, variables, automatic description of the dimensions). All the results can also be exported into a csv file (which can be opened in a spreadsheet such as Excel, for example).

10.3.6 Taking Things Further

1. MCA as a pre-processing method to transform qualitative variables into quantitative variables:

The principal components (coordinates of the individuals or scores) of the MCA are continuous quantitative variables which summarise a set of qualitative variables and which can be used as input of unsupervised classification, for example. In this way, MCA can be seen as a pre-processing step that can convert qualitative variables into quantitative ones. To implement unsupervised classification on the principal components of the MCA, see Worked Example 11.1.

2. Characterising the categories of one specific qualitative variable:

It can also be interesting to use the **catdes** function (see Worked Example 11.1) to describe one specific qualitative variable according to quantitative and/or qualitative variables.

3. Constructing confidence ellipses around categories:

Confidence ellipses can also be drawn around the categories of a categorical variable (i.e. around the barycentre of the individuals carrying that category). These ellipses allow one to visualise whether two categories are significantly different or not. It is possible to construct confidence ellipses for all the categories for a number of categorical variables using the **plotellipses** function:

```
> plotellipses(res.mca,keepvar=c("Loan","Unpaid","Insurance",
      "Profession"))
```

4. Handling missing values in MCA:

It is possible to perform MCA with missing values using the missMDA package. The first step consists of imputing the missing values in the indicator matrix of dummy variables (or disjunctive data table) using the **imputeMCA** function and the second step consists of performing MCA using this completed indicator matrix of dummy variables:

```
> tab.disj.comp <- imputeMCA(my.incomplete.dataset)
> res.mca<-MCA(my.incomplete.dataset, tab.disj=tab.disj.comp)
```

Theoretical reminders on MCA are available Greenacre and Blasius (2006), Greenacre (2007), Lebart et al. (1984) or Govaert (2009) as well as in Husson et al. (2010). The last book also details different case studies using FactoMineR.

11

Clustering

11.1 Ascending Hierarchical Clustering

11.1.1 Objective

Ascending hierarchical clustering (AHC) constructs a hierarchy of individuals that is graphically represented by a hierarchical tree also named a dendrogram. Pruning this tree yields groups (clusters) of individuals. Hierarchical clustering requires to define a distance and an agglomerative criterion. Many distances are available (Manhattan, Euclidean, etc.) as well as several agglomeration methods (Ward, single, centroid, etc.).

To perform AHC, we use the **agnes** function from the cluster package. Using this function, it is possible to have as an input a dissimilarity matrix or a data table individuals × quantitative variables (this is not possible with the **hclust** function). We here present an example in which the individuals are characterised by quantitative variables (the case of qualitative variables is mentioned in Section 11.1.6, "Taking Things Further"). Since we often combine principal component methods and hierarchical clustering to enrich the description of the data, we chose the Euclidean distance and the Ward's agglomerative criterion. Indeed, principal component methods are based on the notion of variance, which is related to Euclidean distance and Ward criterion.

11.1.2 Example

Let us again examine the dataset of the athletes participating in the decathlon (see Worked Example 10.1) and construct an ascending hierarchical clustering of the athletes' performance profiles. We are only interested in the ten performance variables.

11.1.3 Steps

1. Read the data.

2. Standardise the data if necessary.

3. Construct the ascending hierarchical clustering.

4. Prune the hierarchical tree.

5. Characterise the clusters.

11.1.4   Processing the Example

1. Reading the data:

```
> decath <- read.table("decathlon.csv",sep=";",header=TRUE,
    row.names=1)
```

2. Standardising the data if necessary:

Prior to performing an AHC, the variables can be standardised. For the decathlon dataset, we do not have the choice. Standardisation is essential as the variables are of different units. When the variables are of the same units, both solutions are possible and lead to two separate analyses. This decision is therefore crucial. Standardisation means attributing the same importance to each of the variables when calculating the distance between individuals. Not standardising the data means attributing more importance to the variables with a high standard deviation. Here, the data is standardised using the **scale** function.

3. Constructing the ascending hierarchical clustering:

To perform an AHC using the cluster package, the preinstalled package must be loaded (see Section 1.6). We then use the **agnes** function:

```
> library(cluster)
> res.ahc <- agnes(scale(decath[,1:10]), method= "ward")
> plot(res.ahc, which.plots=2, main="Dendrogram",
    xlab="Individual")
```

The **agnes** function takes as an input the (standardised or non-standardised) dataset. The argument `method` specifies the aggregation method and is set here to `Ward`. By default, the argument `metric` is set to the Euclidean distance. The **plot.agnes** function draws the hierarchical tree (or dendrogram) (see Figure 11.1).

4. Pruning the hierarchical tree:

Given the form of the hierarchical tree (Figure 11.1), we can "prune" it to construct clusters of individuals. In order to do so, we must first define the height at which the tree will be cut. For example, a cut height of 10 (Figure 11.1) defines two clusters whereas a height of 8.1 defines four clusters.

Different methods are available to choose the number of clusters. Here, we use the heights at which two clusters join together as a guide. To obtain these heights in a numerical vector, change the object `res.ahc` into a `hclust`-type object and extract its height. As these heights are ranked in ascending order, we inverse this order (**rev**):

Dendrogram

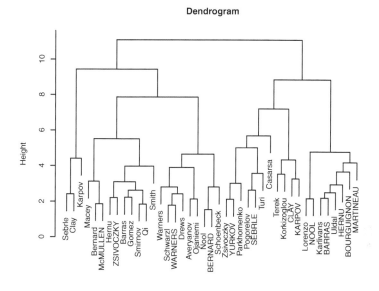

Individual
Agglomerative Coefficient = 0.76

Figure 11.1
Hierarchical tree.

```
> res.ahc2 <- as.hclust(res.ahc)
> plot(rev(res.ahc2$height),type="h",ylab="Height")
```

The height of the fortieth and final bar (Figure 11.2) gives an idea of how difficult it would be to bring together the first individuals and form a clustering of forty clusters. The height of the thirty-ninth bar also gives an idea of the difficulty in changing from this forty-cluster clustering to a thirty-nine-cluster clustering. This change is obtained either by merging two individuals, or by aggregating an individual to an existing cluster. In the same way, the height of the first bar gives an idea of the difficulty in aggregating the two remaining clusters to group the $n = 41$ individuals together.

As we are looking for a small number of homogeneous clusters, we are interested in the heights of the first bars. The most marked jump separates the fifth and sixth bars: it is therefore difficult to go from six to five clusters. One possibility for the cut would therefore be to take six groups. Between the fourth and the third bar there is another less pronounced jump. In order to have enough individuals in each cluster, we chose to use this jump which defines four clusters.

We now need to find out which individuals are in each of these four clusters.

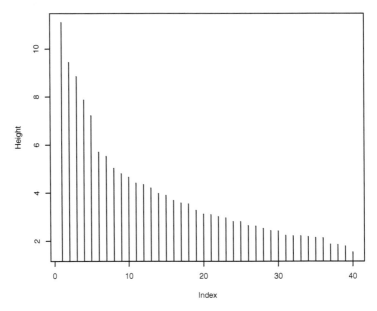

Figure 11.2
Barplot of the heights as a help to choose the number of clusters.

To do this, we use the **cutree** function and indicate the cut, that is, the number of clusters into which the individuals are divided (for example, **k=4**).

```
> clusters.hac <- cutree(res.ahc, k=4)
> clusters.hac
 [1] 1 1 1 2 2 3 2 2 2 2 3 2 2 2 2 2 2 2 2 3 3
[22] 2 3 4 4 3 4 3 3 3 3 2 3 2 2 2 4 4 4 4 4
```

5. Characterising the clusters:

Having grouped the individuals into clusters, it is interesting to describe these clusters in order to interpret the similarities and differences between these groups of individuals. It is possible to describe the clusters using variables with the **catdes** function of the FactoMineR package (see Appendix A.4, p. 269). This function allows one to describe a qualitative variable (more precisely, its categories) from quantitative and qualitative variables. Here, we want to describe the clusters which correspond to the categories of the qualitative variable **clusters.hac**. This variable is thus transformed into a factor. Then, to use the **catdes** function, we construct a unique dataset with the raw data and the cluster variable named **"Cluster"**. The function is then applied on the new dataset, and we specify with the argument **num.var** the index of the variable which will be described.

```
> library(FactoMineR)
> clusters.hac <- as.factor(clusters.hac)
> decath.comp <- cbind.data.frame(decath,clusters.hac)
> colnames(decath.comp)[14] <- "Cluster"
> catdes(decath.comp, num.var = 14)
$quanti
$quanti$'2'
```

	v.test	Mean in category	Overall mean	sd in category	Overall sd
100m	-2.265855	10.89789	10.99805	0.1701572	0.259796
110m.H	-2.397231	14.41579	14.60585	0.3097931	0.466000
400m	-2.579590	49.11632	49.61634	0.5562394	1.139297
1500m	-2.975997	273.18684	279.02488	5.6838942	11.530012

```
$quanti$'3'
```

	v.test	Mean in category	Overall mean	sd in category	Overall sd
1500m	3.899044	290.76364	279.02488	12.627465	11.530012
400m	2.753420	50.43546	49.61634	1.272588	1.139297
Long.	-2.038672	7.09364	7.26000	0.283622	0.312519

The **catdes** function sorts the quantitative variables from the most to the least characteristic as positives (variables for which the individuals of the cluster carry values which are significantly higher than the mean for all the individuals), then from the least to the most characteristic as negatives (variables for which the individuals of the cluster carry values which are significantly lower than the mean for all the individuals). The variables are sorted according to the v-tests defined as

$$\text{v-test} = \frac{\bar{x}_q - \bar{x}}{\sqrt{\frac{s^2}{I_q}\left(\frac{I-I_q}{I-1}\right)}}$$

where $\bar{x}_q$ is the average of variable X for the individuals of cluster q, $\bar{x}$ is the average of X for all the individuals, and I_q is the number of individuals within the cluster q. A v-test with an absolute value of greater than 2 here indicates that the cluster mean is significantly different from the general mean: a positive sign (or negative, respectively) for the v-test indicates that the mean of the cluster is superior (or inferior, respectively) to the general mean. In the example, the results are given for the description of clusters 2 and 3. The individuals from cluster 2 run the 1500 m faster (in a shorter time) than the others. For the individuals in this cluster, the average time taken to run 1500 m is 273.2 s compared with 279.0 s for all the individuals. Three other variables characterise the individuals in this cluster significantly (v-test greater than 2 in absolute value): the 400 m, the 110-m hurdles and the 100 m. These athletes run faster (shorter times) in these events.

For the qualitative variables, it is the categories of the variables which are sorted. These categories are sorted from most to least characteristic when the category is over-represented in the cluster (compared to other clusters) and from least to most characteristic when the category is under-represented in the cluster. In the example, none of the qualitative variables characterise the clusters 2 and 3.

11.1.5   Rcmdr Corner

1. Reading the data from a file:

`Data → Import data → from text file, clipboard, or URL ...`

It must next be specified that the first column contains the individuals' identifiers, which is not possible using the drop-down menu. We therefore need to copy the line of code generated by Rcmdr (top window) and add `row.names=1` to the reading instruction before clicking on `Submit`.

2. Standardising the data if necessary:

`Data → Manage variables in active data set`
`→ Standardize variables...` Then select the variables to be standardised; these variables are added to the raw dataset.

3. Constructing the tree:

`Statistics → Dimensional analysis → Cluster analysis`
`→ Hierarchical cluster analysis...`

We must then select the variables which will enable us to calculate the distances between the individuals (here the standardised variables). Be aware that Rcmdr uses the **hclust** function to conduct the ascending hierarchical clustering. We then need to choose the agglomerative criterion (for example, the Ward's criterion) and the associated distance (with Ward the Euclidean distance squared, which corresponds to the argument of the Euclidean distance for the **agnes** function).

4. Pruning the tree:

`Statistics → Dimensional analysis → Cluster analysis`
`→ Add hierarchical clustering to data set...`

Specify the number of clusters (groups) of individuals to be constructed (here 4). After validation, the dataset contains a supplementary qualitative variable which corresponds to the affiliation to the clusters (by default, the name of the variable is `hclus.label`).

5. Characterising the clusters:

First, load the FactoMineR package and drop-down menu (see Appendix A.4 p. 269). It is thus possible to characterise the clusters of individuals by clicking `FactoMineR → Description of categories`. Then simply choose the variable `hclus.label` to be characterised. By default, all the quantitative and qualitative variables are used to characterize the cluster variable.

11.1.6 Taking Things Further

In this section we describe some extensions of the clustering.

1. Representing the clusters on a principal component graph:

It is possible to colour the individuals on a principal component map according to the cluster with which they are affiliated.

As a reminder, we have performed PCA on the decathlon data with the first ten variables as active (those used to perform the clustering), variables 11 and 12 as quantitative supplementary, and variable 13 as qualitative supplementary. To colour the individuals according to the cluster, we perform a PCA specifying that variable 14 is also qualitative supplementary (variable 14 corresponds to the affiliated cluster). We then construct the graph (**plot.PCA** function) specifying that the individuals are coloured according to variable 14.

```
> res.pca <- PCA(decath.comp, quanti.sup = 11:12,
      quali.sup = 13:14, graph=F)
> plot(res.pca, choix = "ind", habillage=14)
```

2. Clustering from qualitative variables:

If the individuals are described by qualitative variables, there are two possible solutions:

- Construct an appropriate distance or dissimilarity like the Jaccard similarity coefficient, for example (see Kaufman and Rousseeuw, 1990);

- Transform the qualitative variables into quantitative variables using a multiple correspondence analysis (see Worked Example 10.3). Then retrieve the coordinates of the individuals (principal components or scores) and perform the AHC on these coordinates (see below for the example of running together PCA-AHC).

3. Clustering from the principal components:

It can be interesting to perform a principal component method (PCA or MCA) prior to performing a clustering in order to omit the information contained within the last dimensions which can be considered as noise. In the present example, as all data is quantitative, the principal component method is a PCA. We here preserve the information contained within the first eight dimensions using the argument `ncp=8` of the PCA, which represents 96% of the total variance. We then perform the AHC on the coordinates of the individuals:

```
> library(FactoMineR)
> res.pca <- PCA(decath[,1:10], graph=FALSE, ncp=8)
> res.ahc <- agnes(res.pca$ind$coord, method="ward")
```

Traditionally, the aggregation method used when running an AHC on principal components is the Ward's agglomerative criterion (criterion based on variance). The same procedure can be used with MCA.

Remark
When we conduct a clustering on the principal components, the new data table (the coordinates of the individuals) must under no circumstances be standardised or the distances between individuals will be distorted.

4. Consolidating a clustering:

It can be interesting to "consolidate" the clustering with a partitioning method such as k-means (see Worked Example 11.2), specifying the number of clusters to be constructed. In order to do this, we initialise the centres of the clusters with the barycentres of the constructed clusters using ascending hierarchical clustering. We thus calculate the means for each of the variables (centred and standardised as the AHC is conducted on centred and standardised data) for each cluster using the **aggregate** function:

```
> centre.clusters <- aggregate(scale(decath[,1:10]),
      by=list(clusters.hac),FUN=mean)
> centre.clusters
  Group.1     100m  Long.jump  Shot.put  High.jump ...
1       1 -1.526034  1.9279293  1.653179   1.272289 ...
2       2 -0.380781  0.2312184 -0.123492  -0.106361 ...
3       3  0.138758 -0.5257989  0.485427   0.393352 ...
4       4  1.285824 -0.5491438 -0.994111  -0.765361 ...
```

The **kmeans** function takes as its argument the dataset and the number of clusters or the centres of the clusters.

```
> clusters.kmeans <- kmeans(scale(decath[,1:10]),
      centers=centre.clusters[,-1])
> clusters.kmeans$cluster
  Sebrle    Clay   Karpov    Macey   Warners    ...
       1       1        1        1         2    ...
```

Here the argument `centers=centre.clusters[,-1]` initialises the centres of the clusters (the first column of `centre.clusters` is omitted as this column corresponds to the number of the group and must not be used). We retrieve the new partition within the object `cluster`.

5. Clustering with the **HCPC** function:

The four-point process detailed above are integrated in the **HCPC** (for hierarchical clustering on principal components) function of the FactoMineR package. This function combines principal component methods, hierarchical clustering and partitioning, to better describe and visualise the data. The **HCPC** function performs a hierarchical clustering (with the Euclidean distance and the Ward's agglomerative criterion) on the principal components obtained with principal component methods (PCA for quantitative variables and MCA for qualitative ones). The first step consists of performing the principal component method, here a PCA, and then applying the **HCPC** function on the object res.pca:

```
> res.pca <- PCA(decathlon, quanti.sup=11:12, ncp=Inf,
    quali.sup=13,graph=F)
> res.hcpc <- HCPC(res.pca, consol=FALSE)
```

If the argument ncp=Inf, it means that all the principal components are preserved, which amounts to constructing the clustering from the (scaled) raw data. When the argument ncp is less than the number of variables, some principal components are removed which remains to "denoise" the data. The consol=FALSE argument is used to specify that no consolidation (see item 4) is done after pruning the tree. The **HCPC** function yields an interactive graph (Figure 11.3) with the hierarchical tree from which it is possible to choose a cut height by clicking on the tree. An optimal cut height is also suggested by a horizontal line (here at a height of around 0.80). On the same device, another graph is drawn illustrating increases in within-cluster variability from one partition with $k + 1$ clusters and one partition with k clusters.

Remark that the graphs provided by the **agnes** function and the **HCPC** function are different. Indeed, compared to the graph provided by the **agnes** function, the individuals are sorted according to their similarity (i.e. according to their coordinates on the first principal component). It avoids positioning very different individuals side-by-side on the hierarchical tree if they belong to very different clusters. Furthermore, the individuals and the clusters are grouped together according to variance, whereas with the **agnes** function they are grouped according the square root of the variance. Thus, the hierarchical tree for the **HCPC** function tends to be flattened: it is thus more difficult to distinguish between the clusters for the first groupings of individuals but easier for the last groups of clusters (i.e. the top of the hierarchical tree) and it is thus easier to choose the height of the cut.

After having clicked on the graph, the hierarchical tree is represented in three dimensions on the first two principal components (Figure 11.4). Individuals are coloured according to their cluster.

The clusters (obtained after the hierarchical tree has been pruned) are described by the variables, the dimensions of the principal component method and by the individuals using the following commands:

```
> res.hcpc$desc.var
> res.hcpc$desc.axes
> res.hcpc$desc.ind
```

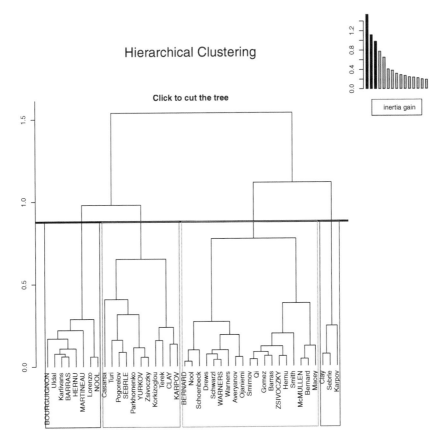

Figure 11.3
Hierarchical tree provided by **HCPC**.

We find here the same description of the clusters as the one given by the **catdes** function. Moreover, the description by the individuals gives, in the object `res.hcpc$desc.ind$para`, the paragon of each cluster (i.e. the individual closest to the centre of each cluster) and its distance to the centre of the cluster.

Reminders of AHC are available in Kaufman and Rousseeuw (1990), Lebart et al. (1984), Hastie et al. (2009, p. 520–528), Murtagh (2005) or Govaert (2009), as well as in Husson et al. (2010). This last book details case studies and the results of a clustering performed after a principal component method.

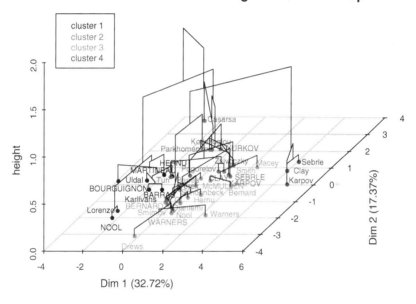

Figure 11.4
Three-dimensional tree on the first two dimensions of the PCA provided by
HCPC.

11.2   The *k*-Means Method

11.2.1   Objective

The *k*-means method provides a partition of individuals, that is, *K* clusters of individuals which are as homogeneous as possible. The partitioning method *k*-means is based on the Euclidean distance between individuals and uses the barycentre of each cluster as a representative individual. *k*-means requires to define the number of clusters *K*. In order to do so, we can, amongst other things, construct an ascending hierarchical clustering and determine a "natural" cut level (see Worked Example 11.1).

The method first consists of choosing (randomly or not) *K* initial cluster centres and then it consists of repeating (until the centres no longer change) the two steps: (1) (re)assign each individual to the cluster to which the individual is closest; (2) update the cluster means, that is, calculate the barycentre of the individuals for each cluster. To perform *k*-means, we use the **kmeans** function. The input of this function is a data table individuals × quantitative variables (the case of qualitative variables is mentioned in Section 11.2.6, "Taking Things Further").

11.2.2   Example

Let us again examine the dataset of the athletes participating in the decathlon (see Worked Example 10.1) and construct a partition of the athletes' performance profiles. We are therefore only interested in the ten performance variables.

11.2.3   Steps

1. Read the data.

2. Standardise the variables if necessary.

3. Construct the partition.

4. Characterise the clusters.

11.2.4   Processing the Example

1. Reading the data:

```
> decath <- read.table("decathlon.csv",sep=";",header=TRUE,
    row.names=1)
```

2. Standardising the variables if necessary:

Prior to performing a k-means algorithm, the variables can be standardised. For the decathlon dataset, we do not have the choice. Standardisation is essential as the variables are of different units. When the variables are of the same units, both solutions are possible and lead to two separate analyses. This decision is therefore crucial. Standardisation means attributing the same importance to each of the variables when calculating the distance between individuals. Not standardising the data means attributing more importance to the variables with a high standard deviation. Here, the data is standardised using the **scale** function.

3. Constructing the partition:

Taking into account the ascending hierarchical clustering performed on this dataset (see Worked Example 11.1), we here partition into $K = 4$ clusters:

```
> results.kmeans <- kmeans(scale(decath[,1:10]), centers=4)
```

The **kmeans** function takes as an input the (standardised or non-standardised) dataset. Then, the argument `centers` can specify either the number of clusters or the centre of each cluster. In the former case, the initialisation of the k-means algorithm is random, that is, four individuals are chosen randomly as starting centre of cluster whereas in the latter case, the centre of each cluster is specified (see item 4 in Section 11.1.6, "Taking Things Further").

```
> results.kmeans
$k$-means clustering with 4 clusters of sizes 11, 13, 5, 12

Cluster means:
          X100m  Long.jump    Shot.put   High.jump      X400m ...
1  1.058138862 -0.8935708 -0.18280501 -0.27095958  1.1885477 ...
2 -0.419569671  0.4278885 -0.04683446 -0.29297229 -0.2955972 ...
3 -1.232016906  1.4285640  1.37419773  1.47464827 -0.9747632 ...
4 -0.002086435 -0.2396743 -0.35427380 -0.04867052 -0.3631204 ...

Clustering vector:
      Sebrle        Clay      Karpov       Macey     Warners   ...
           3           3           3           3           2   ...

Within cluster sum of squares by cluster:
[1] 78.18352 73.25410 28.56075 63.17204
 (between_SS / total_SS =  39.2 %)
```

In output, for each cluster and each variable, the function yields the means of the individuals of the cluster, a vector featuring the cluster number of each individual, and the within-cluster variability.

4. Characterising the clusters:

Having grouped the individuals into clusters, it is interesting to describe these clusters in order to interpret the similarities and differences between these groups of individuals. It is possible to describe the clusters using variables with the **catdes** function of the FactoMineR package (see Appendix A.4, p. 269). This function allows one to describe a qualitative variable (more precisely, its categories) from quantitative and qualitative variables. Here, we want to describe the clusters which correspond to the categories of the qualitative variable `results.kmeans$cluster`. This variable is thus transformed in a factor. Then, to use the **catdes** function, we construct a unique dataset with the raw data and the cluster variable named `"Cluster"`. The function is then applied on the new dataset, and we specify with the argument `num.var` the index of the variable which will be described.

```
> library(FactoMineR)
> decath.comp <- cbind.data.frame(decath,
      factor(results.kmeans$cluster))
> colnames(decath.comp)[14] <- "Cluster"
> catdes(decath.comp, num.var = 14)
$quanti
$quanti$'2'
                        Mean in   Overall      sd in   Overall
              v.test   category      mean   category        sd
Pole.vault  4.452686   5.046154   4.76244   0.176354  0.274589

$quanti$'4'
                        Mean in    Overall     sd in   Overall
              v.test   category       mean  category        sd
  1500m    -3.247660  269.82083  279.02488  5.941016 11.530012
Pole.vault -3.320412   4.538333    4.76244  0.158000  0.274589
```

The **catdes** function sorts the quantitative variables from the most to the least characteristic as positives (variables for which the individuals of the cluster carry values which are significantly higher than the mean for all the individuals), then from the least to the most characteristic as negatives (variables for which the individuals of the cluster carry values which are significantly lower than the mean for all the individuals). The variables are sorted according to the v-tests defined as

$$\text{v-test} = \frac{\bar{x}_q - \bar{x}}{\sqrt{\frac{s^2}{I_q}\left(\frac{I - I_q}{I - 1}\right)}}$$

where $\bar{x}_q$ is the average of variable X for the individuals of cluster q, $\bar{x}$ is the average of X for all the individuals, and I_q is the number of individuals within the cluster q. A v-test with an absolute value of greater than 2 here indicates that the cluster mean is significantly different from the general mean:

a positive sign (or negative, respectively) for the v-test indicates that the mean of the cluster is superior (or inferior, respectively) to the general mean.

In the example, the results are given for the description of clusters 2 and 4. The individuals in cluster 2 jump high in the pole vault. For the individuals in this cluster, the average height is 5.05 m compared with 4.76 m for all the individuals (including those from cluster 2). The individuals in cluster 4 are characterised by the fact that they do not jump as high as the others in the pole vault (v-test less than -2) and that they run the 1500 m faster (in a shorter time) than the others.

For the qualitative variables, it is the categories of the variables which are sorted. These categories are sorted from most to least characteristic when the category is over-represented in the cluster (compared to other clusters) and least to most characteristic when the category is under-represented in the cluster. In the example, none of the qualitative variables characterise the clusters 2 and 4.

11.2.5   Rcmdr Corner

1. Reading the data from a file:

`Data → Import data → from text file, clipboard, or URL ...`

We must specify that the first column contains the individuals' identifiers, which is not possible using the drop-down menu. It is therefore necessary to retrieve the line of code generated by Rcmdr (upper window) and to add `row.names=1` to the reading data from file instruction before clicking on `Submit`.

2. Standardising the variables if necessary:

`Data → Manage variables in active data set`
`→ Standardize variables...` Then select the variables to be standardised; these variables are added to the raw dataset.

3. Constructing the partition:

`Statistics → Dimensional analysis → Cluster analysis`
`→ K-means cluster analysis...`

We must then select the variables which will enable us to calculate the distances between the individuals (here the standardised variables) and specify the number of clusters. It is possible to add the cluster variable to the dataset by specifying `Assign clusters to the data set` and to name the `Assignment variable` (by default, the name is `KMeans`).

4. Characterising the clusters:

First, load the FactoMineR package and drop-down menu (see Appendix A.4, p. 269). It is thus possible to characterise the clusters of individuals by clicking FactoMineR → `Description of categories`. Then simply choose the variable `KMeans` to be characterised. By default, all the quantitative and qualitative variables are used to characterize the cluster variable.

11.2.6   Taking Things Further

1. Representing the clusters on a principal component graph:

It is possible to colour the individuals on a principal component map according to the cluster to which they are affiliated.

As a reminder, we have performed PCA on the decathlon data with the first ten variables as active (those used to perform the clustering), variables 11 and 12 as quantitative supplementary and variable 13 as qualitative supplementary. To colour the individuals according to the cluster, we perform a PCA specifying that variable 14 is also qualitative supplementary (variable 14 corresponds to the affiliated cluster). We then construct the graph (**plot.PCA** function) specifying that the individuals are coloured according to variable 14.

```
> res.pca <- PCA(decath.comp, quanti.sup = 11:12,
      quali.sup = 13:14, graph=F)
> plot(res.pca, choix = "ind", habillage=14)
```

2. Partitioning from qualitative variables:

If the individuals are described by qualitative variables, it is impossible to construct a clustering using the method detailed above. However, as with AHC, it is possible to come back to quantitative data by conducting a prior multiple correspondence analysis (see Worked Example 10.3) and retrieving the coordinates of the individuals on the principal components. Be careful: in that case, data must in no circumstances be standardised or the distances between individuals will be distorted.

3. Other partition methods:

Other partitioning methods are available since the distance between individuals can be non-Euclidean and the representative object of each cluster can be modified. For example, the **pam** function from the cluster package performs a partition around a median individual in each group rather than the barycentre. This function can have as an input a data table or a dissimilarity matrix.

Theoretical reminders on partitioning methods are available in Lebart et al. (1984) and Hastie et al. (2009).

Appendix

A.1   The Most Useful Functions

In this section we offer a non-exhaustive list of the most useful functions in R. These functions are divided under different headings: generic functions, numerical functions, data handling functions, distributions of probabilities, basic statistical functions, advanced statistical functions, graphical functions, reading and exportation functions, text management and finally a few other miscellaneous useful functions. For each function we provide a concise definition and an example of its use.

A.1.1   Generic Functions

In R there are generic functions, that is, functions which can be called by the same order but which yield different results depending on the class of object to which they are applied. The main generic functions are as follows:

Function	Description
print	Writes the results (either all results or an extract)
plot	Constructs a graph
summary	Summarises the results of model fitting functions

There are many functions which begin with **print**, **plot** or **summary**, for example, **print.lm**, **print.PCA**, **print.rpart**, etc. They can all be called using the generic instruction **print** rather than by **print.lm**, **print.PCA**, **print.rpart**, etc. However, to obtain help with a function which writes an rpart object for example, we must write help("print.rpart").

A.1.2   Numerical Functions

Function	Description
abs(x)	Absolute value
sqrt(x)	Square root
ceiling(x)	Gives the smallest following integer: **ceiling**(5.24)= 6, **ceiling**(5)= 5, **ceiling**(-5.24)= −5

Function	Description
floor(x)	Gives the largest previous integer: **floor**(5.24)= 5, **floor**(5)= 5, **floor**(-5.24)= −5
trunc(x)	Truncates the value of x to 0 decimal digits: **trunc**(-5.24)= −5
round(x, digits=n)	Rounds to n decimal digits: **round**(5.236, digits=2)= 5.24
signif(x, digits=n)	Rounds to n total digits: **signif**(5.236, digits=2)= 5.2
cos(x), **sin**(x), **tan**(x), **acos**(x), **cosh**(x), **acosh**(x), etc.	Trigonometric functions
log(x)	Natural logarithm
log10(x)	Base 10 logarithm
exp(x)	Exponential

A.1.3   Data Handling Functions

Function	Description
c	Concatenates within a vector
cbind	Concatenates tables one next to the other (juxtaposition in columns), see Section 2.5
cbind.data.frame	Juxtaposes data-frames into columns, see Section 2.5
rbind	Juxtaposes tables in rows (Caution! This function puts one row on top of another without taking column names into account), see Section 2.5
rbind.data.frame	Juxtaposes data-frames in rows; the column names of the data-frames must be the same (the columns are sorted in the same order for all the tables in order to link the variables prior to concatenation), see Section 2.5
merge	Merges the tables according to a key, see Section 2.5
sort	Sorts vectors in ascending order (or descending order if decreasing = TRUE)
order	Sorts a table according to one or more columns (or rows): x[**order**(x[,3], -x[,6]),] ranks the table x depending on (ascending) the third columns of x then, in the case of equality in the third column of x, depending on (descending) the sixth column of x
by(data, INDICES, FUN)	Applies the FUN function to each level of the vector INDICES in the data table

Function	Description
dimnames	Yields the names of the dimensions of an object (list, matrix, data-frame, etc.)
rownames	Yields the row names of a matrix
row.names	Yields the row names of a data-frame
colnames	Yields the column names of a matrix
col.names	Yields the column names of a data-frame
names	Yields the names of an object (list, matrix, data-frame, etc.)
dim	Yields the dimensions of an object
nrow or **NROW**	Yields the number of rows in a table (in capitals, yields a response even if the object is a vector)
ncol or **NCOL**	Yields the number of columns in a table (in capitals, yields a response even if the object is a vector)
factor	Defines a vector as a factor (if `ordered=TRUE` the levels of the factors are taken to be ordinal)
levels	Yields the levels of a factor
nlevels	Yields the number of levels of a factor
as.data.frame(x)	Converts x into a data-frame
as.matrix(x)	Converts x into a matrix
as.list(x)	Converts x into a list
as.vector(x)	Converts x into a vector
is.data.frame(x)	Tests whether x is a data-frame
is.matrix(x)	Tests whether x is a matrix
is.vector(x)	Tests whether x is a vector
is.list(x)	Tests whether x is a list
class(x)	Yields the class of object x (matrix, data-frame, list, etc.)
mode(x)	Yields the mode of the object x (numeric, logic, etc.)
as.character(x)	Converts x into a character
as.numeric(x)	Converts x into a numeric
as.integer(x)	Converts x into an integer
as.logical(x)	Converts x into a Boolean
is.character(x)	Tests whether x is a chain of characters
is.numeric(x)	Tests whether x is a numeric
is.integer(x)	Tests whether x is an integer
is.logical(x)	Tests whether x is a Boolean

Function	Description
which	Yields the positions of the true values of a vector or a logic table: the parameter `arr.ind=TRUE` yields the numbers of the rows and columns in the table (Section 2.4.1, p. 41): **which**(`c(1,4,3,2,5,3)` == `3`) yields 3 6; **which**(**matrix**(`1:12,nrow=4`) `==3,arr.ind=TRUE`) yields (row 3, column 1)
which.min	Yields the index of the minimum of a vector
which.max	Yields the index of the maximum of a vector
is.na	Tests whether the piece of data is missing
is.null	Tests whether the piece of data is null
length	Yields the length of a list or a vector
any	Tests whether at least one value of a logic vector is true: **any**(**is.na**(`x`)) yields TRUE if at least one piece of data is missing in x
split(`x,fac`)	Divides the table x according to the levels of `fac`

A.1.4   Probability Distributions

There are many probability distributions available in R. All the functions begin with the letters d, p, q or r and end with the name of the distribution.

The letter d indicates that the function calculates the distribution, the letter p indicates that the function calculates a probability, the letter q indicates that the function calculates a quantile, and the letter r indicates that the function generates a random number.

When we generate a random series of numbers it can be interesting to again generate the same series of numbers at a later date. In this case, we must indicate the "random seed generator" used to trigger the series, using the command **set.seed**(`1234`) (the seed can be changed by using a different integer than 1234).

Function	Description
pnorm(q)	Yields the probability $\mathbb{P}(X \leq q)$ for $X \sim \mathcal{N}(0,1)$: **pnorm**(1.96) = 0.975. To calculate $\mathbb{P}(X > q)$ we use the argument `lower.tail = FALSE`
qnorm(p)	Yields the quantile of order p of the $\mathcal{N}(0,1)$ distribution: **qnorm**(0.975) = 1.96
pbinom(q, size, prob)	Yields the probability $\mathbb{P}(X \leq q)$ for binomial distribution $\mathcal{B}$(`size,prob`): **pbinom**(5, 10, .5) = 0.623
ppois(q, lamda)	Yields the probability $\mathbb{P}(X \leq q)$ for X following a Poisson distribution with parameter `lambda`
punif(q, min, max)	Yields the probability $\mathbb{P}(X \leq q)$ for X according to a uniform distribution on $[\text{min}, \text{max}]$

Function	Description
pchisq(q, df)	Yields the probability $\mathbb{P}(X \leq q)$ when X follows a chi-square distribution with df degrees of freedom
pt(q, df)	Yields the probability $\mathbb{P}(X \leq q)$ when X follows a Student's t distribution with df degrees of freedom
pf(q, df1, df2)	Yields the probability $\mathbb{P}(X \leq q)$ when X follows a Fisher's distribution with df1 and df2 degrees of freedom
sample(x, size, replace = FALSE)	This function is used to select a sample of size elements of the vector x without replacement (with replacement if replace=TRUE)
set.seed(n)	Is used to choose a seed for the random number generator; n must be an integer

A.1.5 Basic Statistical Functions

The following statistical functions are used to describe a quantitative variable x. For all these functions, the parameter na.rm=TRUE enables us to eliminate missing data prior to calculation. If na.rm=FALSE and if there is missing data, the function yields an error message.

Function	Description
mean(x, na.rm=TRUE)	Mean of x calculated from the present data
sd(x)	Standard deviation of x
var(x)	Variance of x if x is a vector, or variance-covariance matrix if x is a matrix
cor(x)	Matrix of the correlations of x
median(x)	Median of x
quantile(x, probs)	Quantiles of x for the given probabilities probs
range(x)	Range of x
sum(x)	Sum of the elements of x
min(x)	Minimum of x
max(x)	Maximum of x
sign(x)	Yields the sign of x (positive or negative)
scale(x, center=TRUE, scale=TRUE)	Centres (center=TRUE) and standardises (scale=TRUE) x
colMeans(x)	Calculates the mean for each column in table x
rowMeans(x)	Calculates the mean for each row in table x

Function	Description
apply(x,MARGIN, FUN)	Applies the function FUN to the rows or columns in the table x: **apply**(x, 2, mean) calculates the mean of each column of x; **apply**(x, 1, sum) calculates the sum for each row of x
aggregate(x,by, FUN)	Applies the function FUN to x according to the list of factors offered in by: **aggregate**(x, by =list(vec),mean) calculates the means of x for each level of vec

A.1.6 Advanced Statistical Functions

Function	Description
t.test	Constructs a confidence interval for a mean, tests the equality of a mean to a given value or constructs a test to compare the means of two subsets; **t.test**(x) constructs a confidence interval for the mean of x and tests the equality of the mean of x to the value mu (by default mu=0); **t.test**(x~fac) constructs the test of equality of means in two given sub-populations defined by the factor fac (which must have two categories). By default, the test is constructed with unequal variances var.equal = FALSE; the test is by default bilateral. We can construct a unilateral test using alternative = "less" or alternative = "greater"; see Worked Examples 6.1 and 6.3
var.test	Constructs a test of comparison of variances; see Worked Example 6.3
chisq.test	Constructs a chi-square test; see Worked Example 6.2
prop.test	Constructs a test of equality of proportions; see Worked Examples 6.4 and 6.5
lm(formula)	Constructs a linear model, that is, a multiple regression, an analysis of variance or covariance depending on the nature of the explanatory variables; see Worked Examples 7.1, 7.2, 8.1, 8.2, and 8.3
anova	Yields the analysis of variance table
aov(formula)	Constructs the analysis of variance model defined by formula; if formula =y~x1 + x2 + x1:x2, constructs a model with the main effects x1 and x2 and with the interaction of x1 with x2; see Worked Examples 8.1 and 8.2

Function	Description
glm	Constructs a generalised linear model; see Worked Example 9.2
dist(x)	Constructs a matrix of distances between the rows of the matrix x
PCA	Function of the FactoMineR package which constructs a principal component analysis with the possibility of adding supplementary individuals and quantitative and/or qualitative variables; see Worked Example 10.1
CA	Function of the FactoMineR package which constructs a correspondence analysis; see Worked Example 10.2
MCA	Function of the FactoMineR package which constructs a multiple correspondence analysis with the possibility of adding supplementary individuals and quantitative and/or qualitative variables; see Worked Example 10.3
dimdesc	Function of the FactoMineR package which describes the dimensions of principal component methods
catdes	Function of the FactoMineR package which describes a qualitative variable according to quantitative and/or qualitative variables; see Worked Examples 10.3 and 11.1
agnes	Function which constructs an ascending hierarchical classification; see Worked Example 10.2
kmeans(x, centers)	Constructs a K-means classification from the data table x; centers corresponds to the number of classes or the centres of classes used to initiate the algorithm; see Worked Example 11.2
HCPC	Constructs a classification from the results of a principal component methods; see Section 11.1.6
lda	Function of the MASS package used to conduct a linear discriminant analysis, see Worked Example 9.1
rpart	Function of the rpart package which constructs a regression (or segmentation) tree, see Worked Example 9.3

A.1.7 Graphical Functions

Function	Description
barplot	Yields a bar chart
hist	Yields a histogram

Function	Description
boxplot	Yields a boxplot; **boxplot**(y~fac) yields a graph with a boxplot for each category of the factor fac
pie	Yields a pie chart
points	Plots points on a preexisting graph
lines	Plots lines on a preexisting graph
curve	Draws the curve for a function
abline(a,b)	Draws the line with slope b and intercept a on a preexisting graph
legend	Adds a legend to a preexisting graph; **legend**("topleft", legend = ...) draws the legend at the top left
scatterplot(y~x)	Constructs a scatterplot y according to x; by default, a regression line (reg.line=TRUE) and a non-parametric adjustment curve (by default, smooth=TRUE) are drawn. Boxplots are also provided for x and y (by default, boxplots="xy"). **scatterplot**(y~x|z) constructs a scatterplot for each category of z
pairs	Constructs scatterplots for each pair of variables of a table
persp	Draws graphs in perspective, or response surfaces
image	Constructs three-dimensional response surfaces
locator	Reads the position of the cursor on the graph
identify	Searches for the individuals with the coordinates closest to the position of the cursor
colors()	Provides the list of the 657 colours defined by default in R
graphics.off()	Closes all the graphs which are open
X11()	Creates a new empty graph window
pdf, **postscript**, **jpeg**, **png**, **bmp**	Used to save a graph in pdf, postscript, jpeg, png or bmp format; all the functions are used in the same way: **pdf**("mygraph.pdf"); *graph order*; **dev.off**(); see Section 3.1.5

A.1.8   Import and Export Functions

Function	Description
read.table	Reads a file in table format and creates a data-frame
read.csv	Reads a file with a csv extension containing a table and creates a data-frame
scan	Reads the data originating in a vector or list from a console or file
write	Writes the data in a file

Function	Description
write.table	Writes a table in a file
save	Saves R objects in a file .Rdata
load	Loads objects saved using the save function
history	Retrieves the most recent command lines
savehistory	Saves the command lines in a .Rhistory file
loadhistory	Reads the command lines saved in a .Rhistory file

A.1.9 Text Management

Function	Description
substr(x, start=n1, stop=n2)	Extracts or replaces a sub-chain of characters: **substr**("abcdef", 2, 4) yields "bcd"
grep(pattern, x, fixed=FALSE)	Gives the indices for the elements of list x for which the sub-chain pattern is present. If fixed=FALSE then pattern is an expression, if fixed=TRUE then pattern is a chain of characters
sub(pattern, replacement, x, fixed=FALSE)	Finds, within the chain x, the sub-chain pattern and replaces it with replacement: sub("man","R useR","Hello man") returns "Hello R useR". Replacement only occurs once. See also **gsub** for multiple replacement.
strsplit(x, split)	Cuts a chain of characters into multiple sub-chains according to the character split: **strsplit**("abedtedr","e") returns "ab" "dt" "dr"
paste(..., sep="")	Concatenates multiple chains of characters by separating them with sep
toupper(x)	Writes x in uppercase characters
tolower(X)	Writes X in lowercase letters
apropos(what)	Yields objects containing the chain of characters what in their name

A.1.10 Some Other Useful Functions

Function	Description
seq(from, to, by)	Generates a sequence: **seq**(1,9,2) gives 1 3 5 7 9
rep(x, ntimes)	Repeats sequence x a given number of times: **rep**(4:6,2) gives 4 5 6 4 5 6. It is also possible to repeat each term of the object: **rep**(4:6,each=2) gives 4 4 5 5 6 6

Function	Description
cut(x, n)	Divides a continuous variable into a qualitative variable with **n** levels
solve	Inverts a matrix or resolves a linear system; see Section 1.4.5
eigen	Gives the eigenvalues and eigenvectors of a matrix; see Section 1.4.5
svd	Compute the singular-value decomposition of a rectangular matrix

A.2   Writing a Formula for the Models

Many methods, such as linear regression, analysis of variance, analysis of covariance, logistic regression, etc., require a model to be written. For all these functions, the models are written in similar ways. In this section we show how to do this using a number of examples:

- Y ~ .: model with the variable Y explained by all the other variables in the dataset

- Y ~ x1+x2: model with the variable Y explained by the variables x1, x2; equivalent to Y ~ 1+x1+x2

- Y ~ -1+x1+x2: model with the variable Y explained by the variables x1, x2 without the constant (-1 eliminates the constant)

- Y ~ x1+x2 + x1:x2: model with the variable Y explained by the variables x1, x2 and the interaction between x1 and x2

- Y ~ x1*x2: equivalent to the the previous model

- Y ~ x1*x2*x3: is equivalent to the model with all the main effects and the interactions between x1, x2 and x3, thus Y ~ x1+x2+x3 + x1:x2 + x1:x3 + x2:x3 + x1:x2:x3

- Y ~ (x1+x2):x3: is equivalent to the model Y ~ x1:x3+x2:x3

- Y ~ x1*x2*x3 - x1:x2:x3: is equivalent to the model Y ~ x1+x2+x3 + x1:x2 + x1:x3 + x2:x3

- Y ~ x1+x2 %in% x1: model with the effects of x1 and of x2 organised into a hierarchy (or nested) in x1

- Y ~ sin(x1)+sin(x2): model of Y by sin(x1) and sin(x2)

- Y ~ x1 | fac: model where the effect of x1 is possible for each level of fac

- `Y ~ I(x1^ 2)`: model with the variable `Y` explained by $x1^2$; `I(.)` protects the expression `x1^2`, else `x1^2` is interpreted as `x1*x1 = x1+x1+x1:x1 = x1`

- `Y ~ I(x1+x2)`: model where the variable `Y` is explained by the constant and the variable resulting from the sum (individual by individual) of the variables `x1` and `x2`; `I(.)` protects the expression `x1+x2`, otherwise `x1+x2` is interpreted as two explanatory variables

A.3 The **Rcmdr** Package

The R Commander graphical user interface is available in the Rcmdr package. With this interface, R can be used with an easy-to-use drop-down menu. The educational advantage of this package is that it also provides the lines of code which correspond to the analyses carried out: users can therefore familiarise themselves with programming in R by seeing which functions are used. The Rcmdr interface does not contain all the functions available in R, nor does it feature all the options for the different functions, but the most common functions are programmed and the most classical options are available.

The package only needs to be installed once (see Section 1.6, p. 24). The interface is then loaded using

> library(Rcmdr)

The interface (Figure A.5) opens automatically. This interface features a drop-down menu, a script window and an output window. When the drop-down menu is used, the analysis is launched and the lines of code which were used to generate the analysis are written in the script window.

The simplest way to import data with Rcmdr is to use an Excel file:
Data → Import data → from Excel, Access or dBase data set ...
For a txt or csv file:
Data → Import data → from text file, clipboard, or URL ... Next specify that the field separator is the space and the decimal-point character is ".".
To check that the dataset has been imported successfully:
Statistics → Summaries → Active data set
If we want to read a dataset in `csv` format which contains the individuals' names, it is not possible to specify in the Rcmdr drop-down menu that the first column contains the identifier. We can therefore import the dataset considering the individuals' names as a variable. We thus modify the line of code written in the script window by adding the argument `row.names=1` and then clicking on `Submit`.

Figure A.5
Main window of Rcmdr.

To change the active dataset, simply click on the menu **Data**. If we modify
the active dataset (for example by converting a variable), the dataset must be
refreshed using
Data → Active data set → Refresh active data set

The output window features the code lines in red and the results in blue.
The graphs are drawn in R. At the end of an Rcmdr session, the script window,
and therefore all the instructions, can be saved along with the output file that
is, all the results. R and Rmcdr can be closed simultaneously using **File →
Exit → From Commander and R.**

Remarks
- Writing in the script window of Rcmdr or in the R window amounts to the

same thing. If an instruction is launched from Rcmdr, it is also recognised in R and vice versa. The objects created by Rcmdr can thus be used in R.

- Rcmdr windows may not open correctly, or may be hidden behind other open windows. In this case, if using Windows, right-click on the R icon or on the shortcut used to launch R, then click on `Properties`, and change the target by adding "`--sdi`" after the file access path `Rgui.exe`, which, for example, gives

```
"C:\Program Files\R\R-2.14.1\bin\Rgui.exe" --sdi
```

A.4   The **FactoMineR** Package

A.4.1   The **FactoMineR** Package

FactoMineR is a package dedicated to exploratory multivariate data analysis in a French way. The most common exploratory multivariate data analysis methods are programmed within it: Principal Component Analysis (**PCA** function), Correspondence Analysis (**CA** function), Multiple Correspondence Analysis (**MCA** function). More advanced methods are also available and can be used to take the structure of the variables or individuals into account: Multiple Factor Analysis (**MFA** function) in the sense of Escofier and Pagès, Hierarchical Multiple Factor Analysis (**HMFA** function) or Dual Multiple Factor Analysis (**DMFA** function). The **catdes** function can be used to describe a qualitative variable according to quantitative and/or qualitative variables. The **condes** function can be used to describe a quantitative variable according to quantitative and/or qualitative variables.

In each of these methods, it is also possible to add illustrative elements: supplementary individuals and supplementary quantitative and/or qualitative variables. For each of these analyses, there are many features which help to interpret the data: representation quality, contribution for individuals and for variables. Graphical representations are the core of each of the analyses, and there are many graphical options available in the package: colouring the individuals according to a qualitative variable, representing only those variables which are best projected on the factorial planes, etc.

The package only needs to be installed once (see Section 1.6, p. 24). The package is then loaded using

```
> library(FactoMineR)
```

A website entirely dedicated to the FactoMineR package can be found at `http://factominer.free.fr`. The methods are described and examples are given in detail.

Remark
Another data analysis package is available in R: the ade4 package. Another website dedicated to this package which also provides a great number of detailed examples can be found at `http://pbil.univ-lyon1.fr/ADE-4`.

A.4.2 The Drop-Down Menu

A graphical interface is also available and can be installed within the Rcmdr package interface (see Appendix A.3). There are two ways of loading the FactoMineR package:

- Install the FactoMineR drop-down menu in Rcmdr permanently. To do this, simply write or copy and paste the following line of code into an R window:

```
> source("http://factominer.free.fr/install-facto.r")
```

  For subsequent use of the FactoMineR drop-down menu, simply launch Rcmdr with the command **library(Rcmdr)**, and the drop-down menu will appear by default.

- Install the FactoMineR drop-down menu in Rcmdr for the current session. To do this, install RcmdrPlugin.FactoMineR (see Section 1.6). Then, each time that you want to use the FactoMineR drop-down menu, launch Rcmdr and then click `Tools → Load Rcmdr plug-in(s)` .... Choose the FactoMineR Plug-in from the list, and then relaunch Rcmdr in order to take this new plug-in into account. This option is more complicated, which is why we advise users to opt for the first possibility. An explanation of the use of the drop-down menu can be found in the PCA section, p. 220.

A.5   Answers to the Exercises

A.5.1   Exercises: Chapter 1

Exercise 1.1 (Creating Vectors)
1. The three vectors are created using the **rep** function with the arguments each and times:

```
> vec1 <- rep(1:5,3)
> vec1
 [1] 1 2 3 4 5 1 2 3 4 5 1 2 3 4 5
```

```
> vec2 <- rep(1:5,each=3)
> vec2
 [1] 1 1 1 2 2 2 3 3 3 4 4 4 5 5 5
> vec3 <- rep(1:4,times=(2:5))
> vec3
 [1] 1 1 2 2 2 3 3 3 3 4 4 4 4 4
```

2. The **paste** function concatenates vectors:

```
> vec4 <- paste("A",0:10,")",sep="")
> vec4
 [1] "A0)"  "A1)"  "A2)"  "A3)"  "A4)"  "A5)"
 [7] "A6)"  "A7)"  "A8)"  "A9)"  "A10)"
```

3. The position of the letter q is first calculated. Then the vector of letters and the vector of indices are pasted:

```
> pos.q <- which(letters=="q")
> vec5 <- paste(letters[1:pos.q],1:pos.q,sep="")
> vec5
 [1] "a1"  "b2"  "c3"  "d4"  "e5"  "f6"  "g7"  "h8"  "i9"
[10] "j10" "k11" "l12" "m13" "n14" "o15" "p16" "q17"
```

Exercise 1.2 (Working with NA)

1. The vector is created, and the mean and variance are computed:

```
> set.seed(007)
> vec1 <- runif(100,0,7)
> mean(vec1)
[1] 3.564676
> var(vec1)
[1] 3.94103
```

2. Missing values are allocated at random:

```
> vec2 <- vec1
> ind <- sample(1:100,10)
> vec2[ind] <- NA
> indNA <- which(is.na(vec2))
 [1]  4 19 33 38 40 49 62 71 90 99
```

3. It is necessary to use the argument `na.rm=TRUE` to compute the mean and variance.

```
> mean(vec2)
[1] NA
```

```
> mean(vec2,na.rm=T)
[1] 3.539239
> var(vec2)
[1] NA
> var(vec2,na.rm=T)
[1] 3.879225
```

4. We delete the missing values and find again the mean and variance previously calculated with the argument na.rm=TRUE:

```
> vec3 <- vec2[-indNA]
> mean(vec3)
[1] 3.539239
> var(vec3)
[1] 3.879225
```

5. If the missing values are replaced by the mean of the variable, then the mean is the same as before but the variance is under-estimated:

```
> vec4 <- vec2
> vec4[indNA] <- mean(vec3)
> mean(vec4)
[1] 3.539239
> var(vec4)
[1] 3.487384
```

6. The missing values are replaced by values drawn from a normal distribution with mean and standard deviation of the variable:

```
> vec5 <- vec2
> vec5[indNA] <- rnorm(length(indNA),mean(vec3),sd(vec3))
> mean(vec5)
[1] 3.545158
> var(vec5)
[1] 3.665774
```

7. The missing values are replaced by values drawn from a Uniform distribution from the minimum to the maximum of the observed values:

```
> vec6 <- vec2
> vec6[indNA] <- runif(length(indNA),min(vec3),max(vec3))
> mean(vec6)
[1] 3.726103
> var(vec6)
[1] 3.863905
```

8. The missing values are replaced by values randomly drawn from the observed values:

```
> vec7 <- vec2
> vec7[indNA] <- sample(vec3,10)
> mean(vec7)
[1] 3.477317
> var(vec7)
[1] 3.89452
```

Exercise 1.3 (Creating, Manipulating and Inverting a Matrix)
1. The matrix mat is created prior to attributing the row and column names:

```
> mat <- matrix(c(1,0,3,4,5,5,0,4,5,6,3,4,0,1,3,2),ncol=4)
> rownames(mat) <- paste("row",1:4,sep="-")
> colnames(mat) <- paste("column",1:4)
```

2. The diagonal elements of the matrix mat are obtained as follows:

```
> vec <- diag(mat)
> vec
[1] 1 5 3 2
```

3. The matrix containing the first 2 rows of mat is obtained as follows:

```
> mat1 <- mat[c(1,2),]
> mat1
      column 1 column 2 column 3 column 4
row-1        1        5        5        0
row-2        0        5        6        1
```

4. The matrix containing the last 2 columns of mat is obtained as follows:

```
> mat2 <- mat[,(ncol(mat)-1):ncol(mat)]
> mat2
      column 3 column 4
row-1        5        0
row-2        6        1
row-3        3        3
row-4        4        2
```

5. To calculate the determinant and invert the matrix, simply use the functions **det** and **solve**:

```
> det(mat)
[1] 60
```

```
> solve(mat)
            row-1     row-2       row-3            row-4
column 1     0.5      -0.5   0.1666667  -5.551115e-17
column 2    -0.6       0.4  -0.4666667   5.000000e-01
column 3     0.7      -0.3   0.4333333  -5.000000e-01
column 4    -1.2       0.8  -0.2666667   5.000000e-01
```

Exercise 1.4 (Selecting and Sorting in a Data-Frame)

1. The iris data is loaded and a new dataset is created by selecting only the rows carrying the value "versicolor" for the Species variable:

```
> data(iris)
> iris2 <- iris[iris[,"Species"]=="versicolor", ]
```

2. The dataset is sorted according to the first variable using the **order** function:

```
> iris2[order(iris2[,1],decreasing=TRUE),]
      Sepal.Length Sepal.Width Petal.Length Petal.Width    Species
51             7.0         3.2          4.7         1.4 versicolor
53             6.9         3.1          4.9         1.5 versicolor
77             6.8         2.8          4.8         1.4 versicolor
...
61             5.0         2.0          3.5         1.0 versicolor
94             5.0         2.3          3.3         1.0 versicolor
58             4.9         2.4          3.3         1.0 versicolor
```

Exercise 1.5 (Using the apply Function)

1. To calculate the benchmark statistics, simply use the **summary** function:

```
> library(lattice) # load the package
> data(ethanol)
> summary(ethanol)
      NOx                 C                E
 Min.    :0.370    Min.    : 7.500    Min.    :0.5350
 1st Qu.:0.953     1st Qu.: 8.625     1st Qu.:0.7618
 Median :1.754     Median :12.000     Median :0.9320
 Mean    :1.957    Mean    :12.034    Mean    :0.9265
 3rd Qu.:3.003     3rd Qu.:15.000     3rd Qu.:1.1098
 Max.    :4.028    Max.    :18.000    Max.    :1.2320
```

2. To calculate the quartiles, we can use the **apply** function:

```
> apply(X=ethanol,MARGIN=2,FUN=quantile)
         NOx       C       E
0%    0.3700   7.500 0.53500
25%   0.9530   8.625 0.76175
50%   1.7545  12.000 0.93200
75%   3.0030  15.000 1.10975
100%  4.0280  18.000 1.23200
```

3. The instruction for the previous question by default yields the quartiles. Indeed, since we have not specified the argument `probs` for the **quantile** function, the argument used by default is: `probs=seq(0,0.25,0.5,0.75,1)` (see the help section for the `quantile` function). To obtain deciles, we have to specify `probs=seq(0,1,by=0.1)` as the argument. The help section for the **apply** function indicates the optional arguments via `...`: `optional arguments to 'FUN'`. It is therefore possible to "pass" `probs=seq(0,1,by=0.1)` as an argument to the `FUN=quantile` function:

```
> apply(ethanol,2,quantile,probs=seq(0,1,by=0.1))
        NOx    C      E
0%    0.3700  7.5 0.5350
10%   0.6000  7.5 0.6496
20%   0.8030  7.5 0.7206
30%   1.0138  9.0 0.7977
40%   1.4146  9.0 0.8636
50%   1.7545 12.0 0.9320
60%   2.0994 12.6 1.0104
70%   2.7232 15.0 1.0709
80%   3.3326 15.0 1.1404
90%   3.6329 18.0 1.1920
100%  4.0280 18.0 1.2320
```

Exercise 1.6 (Selection in a Matrix with the apply Function)
1. The matrix containing the columns of `mat` having all values smaller than 6 is obtained as follows:

```
> mat <- matrix(c(1,0,3,4,5,5,0,4,5,6,3,4,0,1,3,2),ncol=4)
> mat3 <- mat[,apply((mat<6),2,all)]
> mat3
     [,1] [,2] [,3]
[1,]    1    5    0
[2,]    0    5    1
[3,]    3    0    3
[4,]    4    4    2
```

2. Because there is only one row which do not contain 0, we have to use `drop=FALSE` such that the output is a matrix and not a vector:

```
> mat4 <- mat[apply((mat>0),1,all),,drop=FALSE]
> mat4
     [,1] [,2] [,3] [,4]
[1,]    4    4    4    2
```

Exercise 1.7 (Using the lapply Function)
1. The MASS package and the dataset `Aids2` are loaded:

```
> library(MASS)    # load the package
> data(Aids2)
> summary(Aids2)
   state         sex            diag              death            status
 NSW  :1780    F:  89    Min.   : 8302    Min.   : 8469    A:1082
 Other: 249    M:2754    1st Qu.:10163    1st Qu.:10672    D:1761
 QLD  : 226              Median :10665    Median :11235
 VIC  : 588              Mean   :10584    Mean   :10990
                         3rd Qu.:11103    3rd Qu.:11504
                         Max.   :11503    Max.   :11504

     T.categ              age
 hs      :2465    Min.   : 0.00
 blood   :  94    1st Qu.:30.00
 hsid    :  72    Median :37.00
 other   :  70    Mean   :37.41
 id      :  48    3rd Qu.:43.00
 haem    :  46    Max.   :82.00
 (Other) :  48
```

2. The function **is.numeric** returns a boolean: TRUE when the object on which it is applied is numeric. We have to apply this function to each column of the data-frame Aids2 and take the negation (operator !). As the data-frame is a list, applying a function to each column is (usually) equivalent to applying a function to each component of the list; this is the scope of the **lapply** function:

```
> ind <- !unlist(lapply(Aids2,is.numeric))
```

3. We just have to select the variables of the data-frame using ind:

```
> Aids2.qual <- Aids2[,ind]
```

4. We use the **levels** function on each element of the data-frame Aids2.qual:

```
> lapply(Aids2.qual,levels)
```

Exercise 1.8 (Levels of the Qualitative Variables in a Subset)
1. The package MASS and the dataset Aids2 are loaded:

```
> library(MASS)    # load the package
> data(Aids2)
```

2. The selection is done as follows:

```
> res <- Aids2[(Aids2[,"sex"]=="M")&(Aids2[,"state"]!="Other"),]
```

Another method consists in using the **subset** function.

3. The summary indicates that the levels are still the same, M and F, but no individuals take the category F.

```
> summary(res)
   state       sex         diag             death        status
 NSW  :1726   F:   0   Min.   : 8302   Min.   : 8469   A: 947
 Other:   0   M:2518   1st Qu.:10155   1st Qu.:10671   D:1571
 QLD  : 217            Median :10662   Median :11220
 VIC  : 575            Mean   :10583   Mean   :10987
                       3rd Qu.:11104   3rd Qu.:11504
                       Max.   :11503   Max.   :11504

    T.categ              age
 hs       :2260   Min.   : 0.00
 hsid     :  68   1st Qu.:30.00
 other    :  55   Median :37.00
 blood    :  54   Mean   :37.36
 haem     :  40   3rd Qu.:43.00
 id       :  21   Max.   :82.00
 (Other)  :  20
```

4. The attributes of sex are:

```
> attributes(res[,"sex"])
$levels
[1] "F" "M"

$class
[1] "factor"
```

5. We transform the sex variable into a character object and print the attributes of the resulting object:

```
> sexc <- as.character(res[,"sex"])
> attributes(sexc)
NULL
```

6. Transform the sexc into a factor:

```
> sexf <- as.factor(sexc)
> attributes(sexf)
$levels
[1] "M"

$class
[1] "factor"
```

7. We select the indices of the non-numeric variables:

```
> ind <- !unlist(lapply(res,is.numeric))
> ind
  state    sex   diag  death status T.categ    age
   TRUE   TRUE  FALSE  FALSE   TRUE    TRUE  FALSE
```

8. We transform the selected variables into character:

```
> res[,ind] <- lapply(res[,ind],as.character)
```

9. We transform the selected variables into factors:

```
> res[,ind] <- lapply(res[,ind],as.factor)
> summary(res)
 state       sex           diag              death          status
NSW:1726   M:2518   Min.   : 8302   Min.   : 8469   A: 947
QLD: 217            1st Qu.:10155   1st Qu.:10671   D:1571
VIC: 575            Median :10662   Median :11220
                    Mean   :10583   Mean   :10987
                    3rd Qu.:11104   3rd Qu.:11504
                    Max.   :11503   Max.   :11504


     T.categ            age
hs       :2260   Min.   : 0.00
hsid     :  68   1st Qu.:30.00
other    :  55   Median :37.00
blood    :  54   Mean   :37.36
haem     :  40   3rd Qu.:43.00
id       :  21   Max.   :82.00
(Other)  :  20
```

A.5.2 Exercises: Chapter 2

Exercise 2.1 (Robust Reading of Data)

1. The **scan** function allows us to read data from file in the form of character vectors. Since the first row contains characters, this type is compulsory. Moreover, two types of decimal are used: ",," and ".".

```
> myvector <- scan("mydata.csv",what="",sep=";")
Read 20 items
> myvector
 [1] "surname"    "height"     "weight"     "feet_size" "sex"
 [6] "tony"       "184"        "80"         "9.5"        "M"
[11] "james"      "175,5"      "78"         "8.5"        "M"
[16] "john"       "158"        "72"         "8"          "M"
```

2. Replace commas by dots:

```
> myvector <- gsub(",",".",myvector)
```

3. Raw data matrix (with row and column names):

```
> mymatrix <- matrix(myvector,nrow=4,ncol=5,byrow=TRUE)
> mymatrix
       [,1]       [,2]      [,3]     [,4]         [,5]
[1,] "surname"  "height"  "weight" "feet_size"  "sex"
[2,] "tony"     "184"     "80"     "9.5"        "M"
[3,] "james"    "175.5"   "78"     "8.5"        "M"
[4,] "john"     "158"     "72"     "8"          "M"
```

4. Retrieval of column and row names:

```
> namecol <- mymatrix[1,-1]
> namecol
[1] "height"   "weight"    "feet_size" "sex"
> namerow <- mymatrix[-1,1]
> namerow
[1] "tony"   "james" "john"
> mymatrix <- mymatrix[-1,-1]
```

5. Data-frame of data with the row and column names:

```
> mydata <- as.data.frame(mymatrix)
> colnames(mydata) <- namecol
> rownames(mydata) <- namerow
> summary(mydata)
  height   weight   feet_size sex
 158   :1   72:1     42:1      M:3
 175.5:1    78:1     43:1
 184   :1   80:1     44:1
```

The first three variables are factors and must therefore be converted to numerics (Section 2.3.1, p. 35):

```
> for (i in c(1,2,3)){
+    mydata[,i] <- as.numeric(as.character(mydata[,i]))
+ }
> summary(mydata)
     height          weight           feet_size      sex
 Min.   :158.0   Min.   :72.00    Min.   :42.0    M:3
 1st Qu.:166.8   1st Qu.:75.00    1st Qu.:42.5
 Median :175.5   Median :78.00    Median :43.0
 Mean   :172.5   Mean   :76.67    Mean   :43.0
 3rd Qu.:179.8   3rd Qu.:79.00    3rd Qu.:43.5
 Max.   :184.0   Max.   :80.00    Max.   :44.0
```

Exercise 2.2 (Reading Data from File)

```
> test1 <- read.table("test1.csv",sep=",",header=TRUE)
> summary(test1)
   CLONE          B              IN               HT19
 1-105 :9   Min.    :1   Min.    :1.000   Min.    : 3.000
 1-41  :9   1st Qu.:1   1st Qu.:2.750   1st Qu.: 7.968
 18-428:9   Median :1   Median :4.500   Median : 9.055
 18-429:5   Mean    :1   Mean    :4.688   Mean    : 8.556
            3rd Qu.:1   3rd Qu.:7.000   3rd Qu.: 9.925
            Max.    :1   Max.    :9.000   Max.    :11.300
      C19            HT29
 Min.    : 5.00   Min.    : 4.50
 1st Qu.:18.75   1st Qu.:11.38
 Median :21.50   Median :12.75
 Mean    :21.75   Mean    :11.91
 3rd Qu.:26.00   3rd Qu.:13.75
 Max.    :34.00   Max.    :15.25


> test1prn <- read.table("test1.prn",header=TRUE)
> summary(test1prn)
   CLONE          B              IN               HT19
 1-105 :9   Min.    :1   Min.    :1.000   Min.    : 3.000
 1-41  :9   1st Qu.:1   1st Qu.:3.000   1st Qu.: 8.150
 18-428:9   Median :1   Median :5.000   Median : 9.110
 18-429:6   Mean    :1   Mean    :4.727   Mean    : 8.595
            3rd Qu.:1   3rd Qu.:7.000   3rd Qu.: 9.900
            Max.    :1   Max.    :9.000   Max.    :11.300
      C19            HT29
 Min.    : 5.00   Min.    : 4.50
 1st Qu.:19.00   1st Qu.:11.50
 Median :22.00   Median :12.75
 Mean    :21.85   Mean    :11.97
 3rd Qu.:26.00   3rd Qu.:13.75
 Max.    :34.00   Max.    :15.25


> test2 <- read.table("test2.csv", sep=";", header=TRUE,
      na.strings="")
> str(test2)
'data.frame':   999 obs. of  6 variables:
 $ CLONE: Factor w/ 87 levels "","1-105","1-41",..: 2 2 2 2 2 2 2 2 2 3 ...
 $ B    : int  1 1 1 1 1 1 1 1 1 1 ...
 $ IN   : int  1 2 3 4 5 6 7 8 9 1 ...
 $ HT19 : num  8.95 9.11 9.85 10.82 9.4 ...
 $ C19  : int  21 21 26 29 22 19 20 20 24 33 ...
 $ HT29 : num  12.5 13 14.2 15 13 ...


> test3 <- read.table("test3.csv", sep=";", header=TRUE,
      na.strings=".")
```

```
> summary(test3)
      CLONE                B                   IN          HT19
 1-105   : 18   Min.    :1.000   Min.    :1   Min.    :    0.89
 1-41    : 18   1st Qu.:1.000   1st Qu.:3   1st Qu.:    7.35
 18-428  : 18   Median :1.000   Median :5   Median :    9.32
 18-429  : 18   Mean    :1.234   Mean    :5   Mean    :   10.47
 18-430  : 18   3rd Qu.:1.000   3rd Qu.:7   3rd Qu.:   10.50
 18-438  : 18   Max.    :2.000   Max.    :9   Max.    :1142.00
 (Other):891                                  NA's    :   15.00
       C19                HT29
 Min.    : 3.00   Min.    : 1.96
 1st Qu.:17.00   1st Qu.:10.75
 Median :23.00   Median :12.88
 Mean    :22.08   Mean    :12.20
 3rd Qu.:28.00   3rd Qu.:14.00
 Max.    :37.00   Max.    :17.50
 NA's    :20.00   NA's    :15.00
```

Exercise 2.3 (Reading Data from File with Date Format)

1. Read the dataset, skipping the first two lines:

```
> ski <- read.table("test4.csv", sep="|", skip=2, header=TRUE,
      row.names=1)
> summary(ski)
       age                gender          first.time.skiing
 Min.    :24.00   Min.    :0.00    1980-05-01:1
 1st Qu.:28.00   1st Qu.:0.00    1982-01-31:1
 Median :32.00   Median :0.00    1992-01-15:1
 Mean    :30.38   Mean    :0.25    2003-03-16:1
 3rd Qu.:33.00   3rd Qu.:0.25    2005-02-26:1
 Max.    :33.00   Max.    :1.00    2006-03-04:1
                                    (Other)    :2
```

2. Use the format `Date` for the last variable. The format `POSIXct` could be interesting if the time was specified.

```
> ski2<-read.table("test4.csv",sep="|",skip=2,header=TRUE,
      row.names=1,colClasses=c("character","numeric",
      "factor","Date"))
> summary(ski2)
       age          gender first.time.skiing
 Min.    :24.00   0:6    Min.    :1980-05-01
 1st Qu.:28.00   1:2    1st Qu.:1989-07-20
 Median :32.00          Median :2004-03-06
 Mean    :30.38          Mean    :1998-06-01
 3rd Qu.:33.00          3rd Qu.:2006-12-02
 Max.    :33.00          Max.    :2009-03-06
```

Exercise 2.4 (Reading Data from File and Merging)
1. Read the datasets:

```
> state1 <- read.table("state1.csv",sep=";",header=TRUE)
> state2 <- read.table("state2.csv",sep=",",header=TRUE)
> state3 <- read.table("state3.csv",row.names=1,header=TRUE)
```

2. Merge by key (common variable `region` for `state1` and `state3`, then common variable `state` for `state2` and the previous table):

```
> state13 <- merge(state1,state3,by="region")
> state123 <- merge(state2,state13,by="state")
```

Exercise 2.5 (Merging and Selection)
1. Open the two files in Excel (or OpenOffice), save them in text format .csv (with OpenOffice, choose field separator ";"). Then read these two files using

```
> fusion1 <- read.table("fusion1.csv",sep=";",header=TRUE)
> summary(fusion1)
      yhat1                yhat3              yhat2
 Min.   :-0.46520   Min.   :-0.6267   Min.   :-2.1563
 1st Qu.:-0.37217   1st Qu.: 0.2906   1st Qu.:-1.1357
 Median :-0.26458   Median : 1.1447   Median :-0.8284
 Mean   :-0.26641   Mean   : 0.9956   Mean   :-0.9200
 3rd Qu.:-0.17349   3rd Qu.: 1.5200   3rd Qu.:-0.6037
 Max.   :-0.03419   Max.   : 2.8496   Max.   :-0.2903
      yhat4              yhat5
 Min.   :0.5374    Min.   :0.07706
 1st Qu.:0.9553    1st Qu.:0.17804
 Median :1.4079    Median :0.25966
 Mean   :1.3332    Mean   :0.27391
 3rd Qu.:1.6578    3rd Qu.:0.37683
 Max.   :2.1196    Max.   :0.48926
> fusion2 <- read.table("fusion2.csv",sep=";",header=TRUE)
> summary(fusion2)
     Rhamnos              Fucos              Arabinos
 Min.   :-0.5348    Min.   :-2.3026   Min.   :-0.77403
 1st Qu.:-0.3513    1st Qu.:-2.0090   1st Qu.:-0.24294
 Median :-0.1010    Median :-1.4524   Median :-0.03542
 Mean   : 0.0925    Mean   :-1.4028   Mean   : 0.15860
 3rd Qu.: 0.5184    3rd Qu.:-0.7409   3rd Qu.: 0.50681
 Max.   : 1.3447    Max.   :-0.2829   Max.   : 1.39122
      Xylos              Mannos
 Min.   :0.4243    Min.   :0.02623
 1st Qu.:0.9477    1st Qu.:0.15794
 Median :1.1553    Median :0.49551
 Mean   :1.1117    Mean   :0.51873
 3rd Qu.:1.3142    3rd Qu.:0.81277
 Max.   :1.7882    Max.   :1.14627
```

2. Retain two columns for each table and create the data-frame:

```
> fusion1 <- fusion1[,c("yhat1","yhat3")]
> fusion2 <- fusion2[,c("Rhamnos","Arabinos")]
> data <- cbind(fusion1,fusion2)
```

3. Create two variables and add these to the data-frame `data`:

```
> yres1 <- data[,"yhat1"]-data[,"Rhamnos"]
> yres2 <- data[,"yhat3"]-data[,"Arabinos"]
> data <- cbind.data.frame(data,yres1,yres2)
> names(data)
[1] "yhat1"   "yhat3"   "Rhamnos"   "Arabinos" "yres1"   "yres2"
```

Exercise 2.6 (Ventilation)
1. Calculate the frequencies:

```
> tabl <- table(Xfac)
> tabl/sum(tabl)
Xfac
   A    B    C    D
0.60 0.20 0.17 0.03
```

2. Display the categories with a sample size of less than 5%:

```
> lev <- levels(Xfac)
> select <- (tabl/sum(tabl))<0.05
> lev[select]
[1] "D"
```

3. Frequencies of the categories without category D:

```
> mod.select <- lev[!select]
> prov <- factor(Xfac[(Xfac%in%mod.select)],levels=mod.select)
> prov <- table(prov)
> freq <- prov/sum(prov)
> freq
prov
        A         B         C
0.6185567 0.2061856 0.1752577
```

4. Select individuals carrying the category D, random draw (ventilation) of their new category. The factor `Xfac` has categories which are described in `levels` but which are no longer represented (i.e. the categories which have been ventilated, here D). The last row updates the list of factor levels.

```
> for (j in lev[select]) {
+    ## sampling in the levels at random with replacement
+    if (length(mod.select)==1) stop("only one category\n") else
+    Xfac[Xfac==j] <- sample(mod.select,sum(Xfac==j),
        replace=T, prob = freq)
+ }
> Xfacvent <- factor(as.character(Xfac))
> Xfacvent
  [1] A A A A A A A A A A A A A A A A A A A A A A A A A A A
 [28] A A A A A A A A A A A A A A A A A A A A A A A A A A A
 [55] A A A A A A B B B B B B B B B B B B B B B B B B B B C
 [82] C C C C C C C C C C C C C C C B A C
```

Exercise 2.7 (Ventilation by Ordinal Factors)

1. Calculate the frequencies:

```
> Xfac <- ordered(c(rep("0-10",1),rep("11-20",3),rep("21-30",5),
    rep("31-40",20),rep("41-50",2),rep("51-60",2),
    rep("61-70",1),rep("71-80",31),rep("+ de 80",20)),
    levels=c("0-10","11-20","21-30","31-40","41-50","51-60",
    "61-70","71-80","+ 80"))
> tabl <- table(Xfac)
> tabl/sum(tabl)
Xfac
      0-10       11-20       21-30       31-40       41-50
0.01176471 0.03529412 0.05882353 0.23529412 0.02352941
      51-60       61-70       71-80        + 80
0.02352941 0.01176471 0.36470588 0.23529412
```

2. Display the categories to be "ventilated":

```
> p <- 0.05
> select <- (tabl/sum(tabl))<p
> lev <- levels(Xfac)
> lev[select]
[1] "0-10" "11-20" "41-50" "51-60" "61-70"
```

3. As long as there is a category with a sample size of less than 5%, it must be ventilated (and deleted from the list of levels): this is the object of the loop while.

```
> while(any((tabl/sum(tabl))<p)) {
+    ## take the first category whose numbers are too low
+    j <- which(((tabl/sum(tabl))<p))[1]
+    K <- length(lev)  # effective of the categories updated
+    ## merge the next or the previous category for the last one
```

```
+    if (j<K) {
+      if ((j>1)&(j<K-1)) {
+        levels(Xfac) <- c(lev[1:(j-1)],paste(lev[j],
+         lev[j+1],sep="."),paste(lev[j],lev[j+1],sep="."),
+         lev[j+2:K]) }
+      if (j==1) {
+        levels(Xfac) <- c(paste(lev[j],lev[j+1],sep="."),
+         paste(lev[j],lev[j+1],sep="."),lev[j+2:K]) }
+      if (j==(K-1))  {
+        levels(Xfac) <- c(lev[1:(j-1)],paste(lev[j],
+         lev[j+1],sep="."),paste(lev[j],lev[j+1],sep=".")) }
+    } else {
+      levels(Xfac) <- c(lev[1:(j-2)],paste(lev[j-1],
+       lev[j],sep="."),paste(lev[j-1],lev[j],sep="."))
+    }
+    tabl <- table(Xfac) ## table updated and ...
+    lev <- levels(Xfac) ## ... categories updated
+ }
> tabl
Xfac
 0-10.11-20.21-30                31-40 41-50.51-60.61-70
                9                  20                   5
            71-80                + 80
               31                  20
```

Exercise 2.8 (Cross-Tabulation → Data Table)

1. A contingency table can simply be constructed by creating the following matrix:

```
> tabl <- matrix(c(2,1,3,0,0,4),2,3)
> colnames(tabl) <- c("Ang","Mer","Tex")
> rownames(tabl) <- c("Low","High")
```

2. The `tabl` matrix is not of table type. There are missing attributes. Also, the opposite simple operation (using **as.data.frame**) is not directly possible. One solution is to add attributes to this matrix. Instead, more simply, we construct this table by hand:

```
> tabmat <- matrix("",length(tabl),3)
> tabmat[,3] <- as.vector(tabl)
> tabmat[,2] <- rep(rownames(tabl),ncol(tabl))
> tabmat[,1] <- rep(colnames(tabl),each=nrow(tabl))
```

3. Implement a data-frame `tabframe`. Column 3, which is converted to a factor using the **data.frame** function, must be converted into a numeric (Section 2.3.1, p. 35). Then, calculate the total sample size and the number of factors:

```
> tabframe <- data.frame(tabmat)
> tabframe[,3] <- as.numeric(as.character(tabframe[,3]))
> tabframe
    X1    X2 X3
1 Ang  Low   2
2 Ang High   1
3 Mer  Low   3
4 Mer High   0
5 Tex  Low   0
6 Tex High   4
> n <- sum(tabframe[,3])
> nbfac <- ncol(tabframe)-1
```

4. Create the **tab** matrix and the counter:

```
> tabcomplete <- matrix("",n,nbfac)
> iter <- 1
```

5. On the first two rows, we operate the loop on all the rows of the **tabframe** table and control it using the sample size (not null). In the new table, we then repeat the allocation of categories, for as many times as required by the sample size (and therefore not the last column of **tabmat** which contains these sample sizes). The row index for this table is managed by the **iter** counter.

```
> for (i in 1:nrow(tabframe)) {
+    if (tabframe[i,3]>0) {
+       for (j in 1:tabframe[i,3]) {
+          tabcomplete[iter,] <- tabmat[i,-ncol(tabframe)]
+          iter <- iter+1
+       }
+    }
+ }
> data.frame(tabcomplete)
       X1      X2
1   Ang     Low
2   Ang     Low
3   Ang    High
4   Mer     Low
5   Mer     Low
6   Mer     Low
7   Tex    High
8   Tex    High
9   Tex    High
10 Tex    High
```

The **tabmat** matrix is used when attributing the rows of **tabcomplete**. It is not possible to allocate to rows of the data-frame directly. Indeed, a data-frame is a list, as is a row of a data-frame, and it is impossible to allocate a list within a row of a matrix which is itself a vector.

A.5.3 Exercises: Chapter 3

Exercise 3.1 (Draw a function)
1. Define a variation grid x and then draw the sine curve:

```
> x <- seq(0,2*pi,length=1000)
> plot(sin(x)~x,type="l")
```

2. To add the title, we use the **title** function (or we could also directly use the `main` argument within the **plot** function):

```
> title("Plot of the sine function")
```

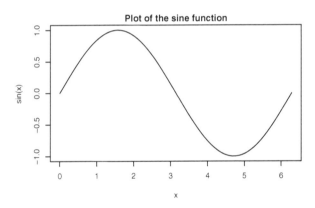

Figure 3.6
Plot of the sine function.

Exercise 3.2 (Comparison of Distributions)
1. To illustrate a normal distribution, simply draw the density using the **dnorm** function. We can then improve the graph by drawing the abscissa axis and then a segment between 0 and the maximum of the normal distribution.

```
> plot(dnorm,-4,4)
> abline(h=0)
> segments(0,0,0,dnorm(0),lty=2)
```

2. To draw new curves we use the **curve** function with the argument add=TRUE. To differenciate between curves, use a different colour for each distribution.

```
> curve(dt(x,5),add=TRUE,col=2)
> curve(dt(x,30),add=TRUE,col=3)
```

3. Simply use the **legend** function and position it at the top left:

```
> legend("topleft",legend=c("normal","Student(5)","Student(30)"),
        col=1:3,lty=1)
```

Exercise 3.3 (Plotting Points)
1. The scatterplot is imported and constructed immediately:

```
> ozone <- read.table("ozone.txt",header=T)
> plot(maxO3~T12,data=ozone)
```

2. To connect the points, simply use `type="l"`; this graph is illegible as we must first sort the data in ascending order on the abscissa axis.

3. The **order** function makes it possible:

```
> sorted <- order(ozone[,"T12"])
> plot(maxO3~T12,data=ozone[sorted,],type="b")
```

Exercise 3.4 (Law of Large Numbers)
1. Create a vector X of length 1000:

```
> set.seed(123)
> X <- rbinom(1000, size=1, prob=0.6)
```

2. The **cumsum** function is used to construct a vector of cumulated sums:

```
> S1 <- cumsum(X)
> M1 <- S1/(1:1000)
> plot(M1, type="l")
> abline(h=0.6, col=2)
```

The resulting graph illustrates the law of large numbers.

Exercise 3.5 (Central Limit Theorem)
1. S_N follows a binomial distribution with parameters N and p, with mean $N \times p$ and standard deviation $\sqrt{N \times p \times (1-p)}$.

2. Set the random seed generator and then simulate a vector of 1000 occurrences of a binomial distribution with the parameters N and p:

```
> set.seed(123)
> p <- 0.5
> N <- 10
```

```
> U10 <- (rbinom(1000, size = N, p=p) - N*p) /sqrt(N*p*(1-p))
> N <- 30
> U30 <- (rbinom(1000, size = N, p=p) - N*p) /sqrt(N*p*(1-p))
> N <- 1000
> U1000 <- (rbinom(1000, size = N, p=p) - N*p)/sqrt(N*p*(1-p))
```

3. Before drawing the curve for the standard normal distribution, create a grid for x varying between -4 and 4. Then, divide the graph window into one row and three columns, plot a histogram and overlay the curve for the normal distribution.

```
> gridx <- seq(-4, 4, by = 0.01)
> par(mfrow=c(1,3))
> hist(U10, xlim=c(-4,4), ylim=c(0,0.6), prob=T)
> lines(gridx, dnorm(gridx), col=4)
> hist(U30, xlim=c(-4,4), ylim=c(0,0.6), prob=T)
> lines(gridx, dnorm(gridx), col=4)
> hist(U1000, xlim=c(-4,4), ylim=c(0,0.6), prob=T)
> lines(gridx, dnorm(gridx), col=4)
```

Exercise 3.6 (Drawing Sunspots)

1. The separator is a comma in the file.

```
> spot<-read.table("sunspots.csv",sep=",",header=T)
> summary(spot)
     nb_spot            year            month             day
 Min.   :  0.00    Min.   :1749    Min.   : 1.00    Min.   :1
 1st Qu.: 15.70    1st Qu.:1807    1st Qu.: 3.75    1st Qu.:1
 Median : 42.00    Median :1866    Median : 6.50    Median :1
 Mean   : 51.27    Mean   :1866    Mean   : 6.50    Mean   :1
 3rd Qu.: 74.92    3rd Qu.:1925    3rd Qu.: 9.25    3rd Qu.:1
 Max.   :253.80    Max.   :1983    Max.   :12.00    Max.   :1
```

2. Create the qualitative variable `thirty`:

```
> thirty <- floor((spot[,2]-1749+30)%/%30)
> thirty <- factor(thirty)
```

3. Check that the colours mentioned do indeed feature in the colour palette.

```
> color<-c("yellow","magenta","orange","cyan","grey","red",
    "green","blue")
> all(color%in%colors())
[1] TRUE
```

4. To draw the chronological series of Figure 3.34, we first construct the graph without a curve and without points (argument `type="n"`). This makes it possible to define the variation ranges of x and y along with the axes' labels. We thus draw parts of the curve one by one, changing the colour for each category of `thirty`:

```
> palette(color)
> coordx <- seq(along=spot[,1])
> plot(coordx,spot[,1],xlab="Time",ylab="Number of sunspots",
     type="n")
> for (i in levels(thirty)) {
>     select <- thirty==i
>     lines(coordx[select],spot[select,1],col=i)
> }
```

Exercise 3.7 (Plotting a Density)
1. To plot the curve for the normal distribution, first define the variation interval of x:

```
> x <- seq(-3.5,3.5,length=1000)
> plot(x,dnorm(x),type="l",ylab="Density")
```

2. To draw a horizontal line, use **abline** and the argument h:

```
> abline(h=0)
```

3–5 For questions 3 to 5, use the functions **polygon**, **arrows** and **text**. In order to be able to write mathematics using the **text** function, use **expression**:

```
> select <- x>=qnorm(0.95)
> absci <- c(x[select],rev(x[select]))
> ordon <- c(rep(0,sum(select)),rev(dnorm(x[select])))
> polygon(absci,ordon,col="blue")
> arrows(2.7,0.2,2,dnorm(2),len=0.1)
> text(2.7,0.2,expression(paste(alpha==5,"%")),pos=3)
```

Exercise 3.8 (Multiple Graphs)
1. To generate the graphs in Figure 3.36, we need to redefine the margins of each graph using the **par** function. We also use the **layout** function to define the layout of the three graphs:

```
> par(mar=c(2.3,2,0.5,0.3))
> layout(matrix(c(1,1,2,3), 2, 2, byrow = TRUE))
> plot(1:10,10:1,pch=0)
> plot(rep(1,4),type="l")
> plot(c(2,3,-1,0),type="b")
```

2. The argument `widths` of **layout** is used to define the width of each column:

```
> par(mar=c(2.3,2,0.5,0.3))
> layout(matrix(c(1,1,2,3), 2, 2, byrow = TRUE),widths=c(4,1))
> plot(1:10,10:1,pch=0)
> plot(rep(1,4),type="l")
> plot(c(2,3,-1,0),type="b")
```

Exercise 3.9 (Plotting Points and Rows)
1. Read data from a file and check reading:

```
> ozone <- read.table("ozone.txt",header=T)
> summary(ozone)
```

2. Draw the graph using the **xyplot** function of the lattice package:

```
> library(lattice)
> xyplot(maxO3~T12,data=ozone)
```

3. Draw the graph and connect the points using the argument `type=c("p","l")` (the graph looks disorganised as the data is not sorted):

```
> xyplot(maxO3~T12,data=ozone,type=c("p","l"))
```

4. Draw the graph using the argument `type="a"` which sorts the abscissa:

```
> xyplot(maxO3~T12,data=ozone,type="a")
```

The curve does not go through all the points, as the function smoothes the curve before drawing it.

Exercise 3.10 (Panel)
1. Read the data and construct a data-frame (denoted `data`) with the first three rows of `maxO3` and T12:

```
> ozone <- read.table("ozone.txt",header=T)
> data <- ozone[1:3,c("T12","maxO3")]
```

2. Repeat the same data six times using a loop:

```
> prov <- data
> for (i in 1:5) prov <- rbind(prov,data)
```

3. Construct the ordinal factor, specifying the order of the levels using `levels`:

```
> type <- ordered(c(rep("h",3),rep("b",3),rep("p",3),rep("l",3),
        rep("s",3),rep("S",3)),levels=c("p","l","b","h","s","S"))
```

4. Add this ordinal factor to the `prov` data: here, a data-frame must be created as some variables are quantitative and others qualitative.

prov <- **cbind.data.frame**(prov,type)

5. Reproduce the graph in Figure 3.10 using **xyplot** and the appropriate panel function:

```
> mypanel <- function(x,y,subscripts) {
+    panel.xyplot(x,y,type=unique(type[subscripts]))
+ }
> xyplot(maxO3~T12|type,data=prov,panel=mypanel,xlab="",ylab="",
    scales=list(draw=FALSE),layout=c(6,1))
```

6. We leave the reader to reproduce this graph using the **factor** function rather than the **ordered** function in question 3.

A.5.4 Exercises: Chapter 4

Exercise 4.1 (Factorial)
1. Construction of the factorial function using the **prod** function:

```
> my.factorial <- function(n) {
+    if (n<0) stop("the integer must be positive")
+    if (n==0) return(1)
+    if (floor(n)!=n) warning(paste(n,"rounded to",floor(n)))
+    result <- prod(1:n)
+    return(result)
+ }
```

Here are three examples of the use of this function:

```
> my.factorial(4)
[1] 24
> my.factorial(4.2)
[1] 24
Message d'avis :
In my.factorial(4.2) : 4.2 rounded to 4
> my.factorial(-3)
Erreur dans my.factorial(-3) : the integer must be positive
```

2. Construction of the factorial function using a loop `for`:

```
> my.factorial <- function(n) {
+    if (n<0) stop("the integer must be positive")
+    if (n==0) return(1)
```

```
+   if (floor(n)!=n){
+       warning(paste(n,"rounded to",floor(n)))
+       n <- floor(n)
+   }
+   result <- 1
+   for (i in 1:n) result <- result*i
+   return(result)
+ }
```

Exercise 4.2 (Ventilation)

1. Construct the ventilation function:

```
> ventilation <- function(Xfac,p=0.05) {
+   if (!is.factor(Xfac)) stop("Xfac must be a factor \n")
+   categories <- levels(Xfac)
+   if (length(categories)<=1) stop("not enough levels \n")
+   tabl <- table(Xfac)
+   select <- (tabl/sum(tabl))<p
+   if (!any(select)) return(Xfac) else {
+     sel <- categories[!select]
+     prov <- factor(Xfac[(Xfac%in%sel)],levels=sel)
+     prov <- table(prov)
+     freq <- prov/sum(prov)
+     for (j in categories[select]) {
+       ## sampling in the levels at random with replacement
+       if (length(sel)==1){
+         warning("1 level\n")
+         Xfac[Xfac==j]<-sel
+       } else Xfac[Xfac==j]<-sample(sel,sum(Xfac==j),
+                   replace=T,prob=freq)
+     }
+     Xfacvent <- factor(as.character(Xfac))
+   }
+   return(Xfacvent)
+ }
```

2. We apply the function from the previous question to each column of the table which is a factor:

```
> ventil.tab <- function (tab, threshold=0.05) {
+ for (i in 1:ncol(tab)) {
+   if (is.factor(tab[,i])) tab[,i]<-ventilation(tab[,i],p=threshold)
+ }
+ return(tab)
+ }
```

Exercise 4.3 (Ventilation on an Ordered Factor)

1. The function is almost or exactly the same as that seen in the answer for Exercise 2.7. To the output we add (**return**) and a few verifications:

```
> ventilation.y <- function(Xfac,p=0.05) {
+ if (!is.ordered(Xfac)) stop("Xfac must be ordered \n")
+ categories <- levels(Xfac)
+ if (length(categories)<=1) stop("not enough levels \n")
+ tabl <- table(Xfac)
+ selecti <- (tabl/sum(tabl))<p
+ if (!any(selecti)) return(Xfac) else {
+   numero <- which(selecti)
+   while(any((tabl/sum(tabl))<p)) {
.. paste here the inside of the while loop from Exercise 2.7
+   }
+ }
+ return(Xfac)
+ }
```

2. We apply the function from the previous question to each column of the table which is an ordinal factor:

```
> ventil.y.tab <- function (tab, threshold=0.05) {
+  for (i in 1:ncol(tab)) {
+    if (is.ordered(tab[,i])) tab[,i]<-ventilation.y(tab[,i],threshold)
+  }
+  return(tab)
+ }
```

Bibliography

Breiman L., Friedman J., Olshen R. and Stone C. (1984). *Classification and Regression Trees*. Wadsworth & Brooks, Monterey, CA.

Clarke G.M. and Cooke D. (2004). *A Basic Course in Statistics*. Wiley, New York, 5 ed.

Collett D. (2003). *Modelling Binary Data*. Chapman & Hall/CRC, Boca Raton, FL, 2 ed.

Devroye L., Györfi L. and Lugosi G. (1996). *A Probabilistic Theory of Pattern Recognition*. Springer, New York.

Faraway J.J. (2005). *Linear Models with R*. Chapman & Hall/CRC, London.

Fox J. and Weisberg S.H. (2011). *An R Companion to Applied Regression*. Sage Publications, London, 2 ed.

Govaert G. (2009). *Data Analysis*. Wiley, New York.

Greenacre M. (2007). *Correspondence Analysis in Practice*. Chapman & Hall/CRC, Boca Raton, FL.

Greenacre M. and Blasius J. (2006). *Multiple Correspondence Analysis and related methods*. Chapman & Hall/CRC, Boca Raton, FL.

Hastie T., Tibshirani R. and Friedman J. (2009). *The Elements of Statistical Learning: Data Mining, Inference, and Prediction*. Springer, New York, 2 ed.

Hosmer D. and Lemeshow S. (2000). *Applied Logistic Regression*. Wiley, New York, 2 ed.

Husson F., Lê S. and Pagès J. (2010). *Exploratory Multivariate Analysis by Example Using R*. Chapman & Hall/CRC, London.

Jolliffe I. (2002). *Principal Component Analysis*. Springer, New York.

Kaufman L. and Rousseeuw P.J. (1990). *Finding Groups in Data: An Introduction to Cluster Analysis*. Wiley, New York.

Lebart L., Morineau A. and Warwick K.M. (1984). *Multivariate Descriptive Statistical Analysis: Correspondence Analysis and Related Techniques for Large Matrices.* Wiley series in probability and mathematical statistics: Applied probability and statistics. Wiley, New York.

McLachlan G. (2004). *Discriminant Analysis and Statistical Pattern Recognition.* Wiley, New York, 2 ed.

Moore D.S., McCabe G.P. and Craig B. (2007). *Introduction to the Practice of Statistics.* Freeman, W. H. & Company, New Zealand, 6 ed.

Murrel P. (2005). *R Graphics.* Chapman & Hall/CRC, New Zealand.

Murtagh F. (2005). *Correspondence Analysis and Data Coding with R and Java.* Chapman & Hall/CRC Press, London.

Sahai H. and Ageel M.I. (2000). *The Analysis of Variance: Fixed, Random and Mixed Models.* Birkhäuser Publisher, Boston, MA.

Varmuza K. and Filmozer P. (2009). *Introduction to Multivariate Statistical Analysis in Chemometrics.* CRC Press, Boca Raton, FL.

Vinzi V., Chin W., Henseler J. and Wang H.e. (2010). *Handbook of Partial Least Squares.* Springer, Boston, MA.

Welch B.L. (1951). On the comparison of several mean values: an alternative approach. *Biometrika*, **38**, 330–336.

Index of the Functions

!= 11, 98
+ 4
-> 6
: 9
<- 6
<= 11
< 11
== 11, 13
= 6
>= 11
> 4, 11
? 5
Ctrl + c 4
Esc 4
Inf 9
NA 8
NaN 9
[ ] 12, 14
[[ ]] 20
.... 6
$ 24
%*% 16
& 12

abline 63, 264
abs 13, 257
aggregate 93, 262
agnes 241, 242, 263
all 11
anova 160, 195, 262
any 11, 41, 260
aov 160, 262
apply 90, 262
apropos........................265
array.......................18, 90
as.character 8, 259

as.complex 8
as.data.frame 22, 259
as.integer 259
as.list.......................259
as.logical 8, 259
as.matrix 14, 259
as.numeric 8, 259
as.vector 259, 285
axis 56, 70

barchart 76
barplot...............76, 114, 263
bartlett.test 124, 172
binom.test 125
boxplot 42, 76, 119, 168, 264
bwplot.........................76
by 94, 95, 258

c 9, 21, 258
CA 222, 263
catdes.........245, 254, 263, 269
cbind...............18, 45, 258
cbind.data.frame 46, 258
ceiling 257
chisq.test 115, 262
chol...........................17
class.........................259
cloud.......................65, 76
col.names 259
colMeans 90, 261
colnames..................44, 259
colors.....................70, 264
contour 64, 76
contourplot 76
coplot....................72, 176
cor 261

cos 258
crossprod 16
cumsum 288
curve 264, 287
cut 36, 266

data 6
data.frame 22
demo 51
density 59, 76
densityplot 76
det 16
dev.off 66
diag 16
dim 259
dimdesc 216, 237, 263
dimnames 13, 21, 22, 259
dist 263
dnorm 287

eigen 16, 266
exp 258

factor 19, 38, 259
floor 258
friedman.test 172

getwd 4
glm 191, 263
graphics.off 264
grep 265

HCPC 248, 263
help 5
help.start 5
hist 59, 73, 263
history 265

identify 43, 264
image 64, 73, 76, 264
interaction.plot 73, 168
is.character 8, 259
is.complex 8
is.data.frame 259
is.integer 259
is.list 259

is.logical 8, 259
is.matrix 259
is.na 9, 13, 260
is.null 8, 260
is.numeric 8, 259
is.vector 259

kmeans 252, 253, 263
kruskal.test 122

lapply 21, 92
layout 68
lda 183, 263
legend . 71, 138, 150, 175, 264, 288
length 8, 260
levelplot 76
levels 19, 37, 259
library 24
lines 63, 264
lm 160, 262
load 4, 265
loadhistory 265
locator 264
log 258
log10 258

matplot 73
matrix 13
max 261
MCA 232, 263
mean 261
median 261
merge 45, 258
min 261
mode 7, 259

names 20, 259
ncol 259
nlevels 19, 259
nrow 259

order 258, 274
ordered 19
outer 64, 96

pairs 73, 76, 264

palette 70
par 67, 70, 71
paste 10, 265
pbinom 260
PCA 209, 263
pchisq 261
pdf 66, 67, 264
persp 64, 73, 76, 264
pf 261
pie 60, 73, 264
plot . 52–59, 63, 70–71, 73, 76, 257
plot3d 66
pnorm 260
points 264
power.anova.test 131
power.t.test 129
ppois 260
print 7, 257
prop.table 115
prop.test 125, 262
ps 67
pt 261
punif 260

q 4
qnorm 260
qqnorm 73
qqplot 73
qr 17
quantile 36, 74, 261, 275
quartz 67

range 261
rbind 18, 43, 258
rbind.data.frame 258
read.csv 264
read.table 7, 30, 264
relevel 38
rep 10, 265, 285
replicate 96
return 97
rm 7
round 115, 258
row.names 259
rowMeans 90, 261

rownames 32, 44, 259
rpart 263
rug 148

sample 90, 261
save 265
save.image 4
savehistory 265
scale 94, 261
scan 10, 264, 278
scatterplot 264
sd 261
seq 9, 88, 265
set.seed 23, 261
setwd 4
shapiro.test 121
sign 261
signif 258
solve 16, 17, 266
sort 258
spineplot 56, 73
split 260
splom 76
sqrt 257
step 195
str 280
strsplit 265
sub 265
substr 11, 265
sum 261
summary 30, 257
svd 16, 266
svg 67
sweep 93

t.test 122, 262
table 47
tapply 92
tolower 265
toupper 265
trellis.device 81
trellis.par.set 80
trunc 258
TukeyHSD 165

unlist 21

update.packages 25

var 261
var.test 122, 262

which 13, 43, 260
which.max 260
which.min 13, 260
wilcox.test 122
wireframe 76
write 264
write.csv 33
write.infile 34
write.table 33, 265

X11 . . . . 67, 69, 264
xfig 67
xtabs 47, 117
xyplot 74, 76–179

Index

A

Add to a graph
    arrows 60
    circles, squares, etc. 60
    diagonal 63
    points 61
    polygon 60
    segments 60
AHC 241–250
    describing the clusters 244
Analysis of covariance 173
Analysis of variance
    change of reference 163
    constraints 171
    contrasts 164, 171
    Fisher's exact test 160
    multi-way 166
    reference level 162
    residual analysis 161
    residuals 161
    Student t-test 162
    studentised residuals 161
    table 160
    two-way 166
    with interaction 166
ANOVA .. *see* Analysis of variance
Argument
    default value 23
    of a function 23
Atomic object 9
Attribute 8
Axis
    size 71

B

Bar chart 54

Boxplot 42, 54, 119, 168

C

CA 222–229
    column coordinate 227
    column representation quality
        227
    contribution to a column . 227
    drop-down menu 227, 228
    eigenvalues 224
    percentage of variance 224
    representation quality of a row
        226
    row contribution 226
    row coordinate 226
    supplementary column ... 224
    with Rcmdr 227
Category .... *see* Level
Circular representation 60
Classification
    tree 199
Close a session 4
Clustering
    AHC 241
    ascending hierarchical 241–250
    consolidation of a partition 248
    dendrogram 242
    describing the clusters ... 244,
        254
    from qualitative variables 247
    hierarchical tree 242
    k-means 252
    on qualitative variables ... 256
    pruning the tree 242

representation of clusters on a
principal component map
247, 256
Colour
    background 70
    choice 70
    foreground 70
    name...................... 70
    numbering 70
    palette 70
    rgb........................ 70
Column
    modify the names 22
    number of................. 17
Commander (package) 267
Concatenation 18
    by column................. 43
    by row 43
Confidence
    ellipse.................... 240
    interval
        of a mean.............. 109
        of a proportion 126
        of a variance ratio 122
Constraints................... 167
Contrasts
    modify with Rcmdr 164
Conversion
    logical to numeric 11
    qualitative to quantitative. 35
    quantitative to qualitative. 34
Convert an object.............. 8
Correspondence Analysis.. *see* CA
CRAN 3
Cross-tabulation 46

D

Data
    missing..................... 8
    outlier (find) 42
    summary................. 30
Data cube 18
Data for worked examples 103
Data-frame 22
    conversion to matrix 23

creation 22
    extraction 22
Date
    reading................... 49
Decomposition
    Cholesky 17
    singular values 16
Device
    definition................. 66
    export 66
    lattice.................... 81
    list 67
Diagonal
    add 63
    dotted 61, 63, 72
    draw 63
    thickness 72
Diagonalisation of a matrix 16
Diagram bar 58, 60
Dimension 17
Distribution box...... *see* Boxplot
Division into classes....... 35, 149
Draw
    arrows 60
    circles, squares, etc. 60
    function 52, 63
    lines...................... 57
    points 54, 57, 134
    polygon 60
    segments 60
Drop-down menu 267

E

Equation systems 17
Estimation of
    coefficients 136
        in ANOVA............. 162
        in logistic regression ... 192
    density
        histogram.............. 59
        kernel.................. 59
Export
    graphs 66
    results 33

F

Factor
 category . . . . *see* level
 combine levels 37
 contrasts 164, 171
 conversion to numeric . . 20, 35
 creation 35
 definition 18
 level
  combine 37
  definition 18
  delete 38
  number 19
  order (change) 37
  rename 37
 ordinal 18
 ventilation 39, 49
Function
 argument 23
 call 23, 98
 creation 97–99
 graph 72, 73
 graphical 76
 result 24, 97
Function library . . . . . . *see* Package

G

Graph(s)
 3D
  function 64
  image 64
  level lines 64
  scatterplot 65–66
 axes
  creation 56, 70
  legend 54, 56
 bar 54, 60
 conditional . . . . *see* Lattice
 diagonal 63
 function 63
 interactive 82
 legend 71, 138, 150, 175
 limits 63, 72
 lines 57, 63
 multiples

  conditional . . . . . *see* Lattice
  on one page 67
 orthonormal 70
 parameters 67, 70, 71
 redefine the margins 291
 scatterplot 52–58
 symbols 57
Graphical symbol 57

H

Help
 on CRAN 6
 online 5, 6
Histogram 59

I

Identifier 29
Identify a point on a graph 43
Identity matrix 16
If then else 89
Image 4
 load 4
 save 4
Individual
 import the name 29
Installing R 3
Interaction
 graphical representation . . 168
 qualitative variables 166
Inversion of a matrix 16

J

Juxtapose tables
 by column 43
 by row 43

K

k-means 252

L

Lattice
 argument `type` 85
 colour 78, 81
 definition 75
 export 81
 font size 80

formulae 76
graphical functions 76
layout 81
legend 78
options 80
panel 74, 79
prepanel 79
scatterplot 74, 179
symbol size 80
themes 81
Legend 71, 138, 150, 175
Level
    combine 37
    definition 18
    delete 38
    number 19
    order (change) 37
    rename 37
Line
    of points 52
    regression 136
Line (drawn)
    dotted 63, 72
    thickness 72
    type 57
Linear algebra 16
Linear regression
    adjusted R^2 136
    BIC criterion 143
    choosing the variables . . . 143,
        146
    coefficient R^2 136
    confidence interval of a predic-
        tion 138
    estimation of parameters . 135
    Mallows' Cp 143
    model without constant . . 136
    multiple 140
    prediction 138, 145
    regression line 136
    residual variance 136
    residuals 137
    simple 133
    studentised residuals 137, 144
List 20

creation of 20
extraction of 20
Loop
    for 87
    repeat 88
    while 88

M
Margin of graph
    redefine 290
Matrix 13
    Cholesky decomposition . . 17
    conversion to data-frame . . 23
    creation 13
    determinant 16
    diagonal 16
    diagonalisation 16
    identity 16
    inversion 16
    product 16
    selection in 14
    singular value
        decomposition 16
    transpose 16
    variance-covariance 261
MCA 230–240
    automatic description of the
        dimensions 237
    contribution of a category 236
    contribution of a variable . 236
    coordinates of a category . 236
    description of a variable . . 240
    eigenvalues 233
    graph of categories 235
    graph of individuals 234
    grouping the categories . . . 232
    percentage of variance 233
    quality of representation of a
        category 236
    score 236
    supplementary qualitative
        variable 232
    supplementary quantitative
        variable 232
    ventilation 232

with Rcmdr 239
Mean
    confidence interval 109
    estimation 110
    test of equality 118
Merge . . . . *see* concatenate
    according to a key 45
Missing value
    definition 8
    reading 30
    search/replace 39
Mode of an object 7
Multiple Correspondence Analysis
        see MCA
Multiple test 165

N

Name 8
    column 44
    individuals 29
    list 20
    matrix 13, 22
    row 32, 44
Normal QQ plot 121
null 7

O

Object 6
    Boolean 7
    chain of characters 7
    change mode 8
    complex number 7
    create 6
    delete 6, 7
    display 6
    mode 7
    real number 7
Outlier (find) 42

P

Package 24
    cluster 242
    FactoMineR 211, 224, 232, 269
    lattice 65
    MASS 183
    missMDA 220, 240

Rcmdr 267
RMySQL 31
rpart 199
tree 199
installation 24
loading 25
update 25
use 25
Partition . . . . *see* *k*-means
Partitioning method . *see* *k*-means
PCA 209–220
    automatic description of the
        dimensions 216
    axes 3, 4, etc. 215
    colouring the points 215
    contribution of a variable . 215
    contribution of an
        individual 215
    coordinates of an
        individual 215
    drop-down menu 218, 239
    eigenvalues 211
    loadings 215
    non-standardised 211
    percentage of variance 211
    principal component 215
    qualitative supplementary
        variable 211
    quality of representation of an
        individual 215
    quantitative supplementary
        variable 211
    representation quality of a
        variable 215
    score 215
    standardised 211
    with Rcmdr 218
Pie chart 60
PLS 147
PLS regression
    residuals 151
Points
    add 61
    cloud 52, 54
    line 57

Power of a test 129
Principal Component Analysis *see*
    PCA
Product of matrices............16
Programming
    condition.................89
    loop.....................87
Proportion
    confidence interval 126
    conformity test...........125
    equality test 127
    test of equality...........128

R

R Commander (package)......267
R function 23
Reading
    data 29–33
    problems.................31
Reduce 94
Regression...*see* Linear regression
    PLS.....................147
    tree.....................199
Rounding.....................115
Rows (of an object)
    modify the names 22
    number of................17

S

Sampling 90, 261
Saving objects in an image......4
Scatterplot 52–58, 134
Segmentation....*see* Decision tree
Selection
    of a vector 12
Session
    Linux 4
    Mac......................5
    Windows..................4
Size..........................71
    axes.....................71
    fonts 60, 71
    points 57, 71
    title....................71
Sorting a data table 274

Standard deviation
    estimation 110
Standardise...................94
Studentised residuals 144

T

Table of analysis of variance .. 160
Test
    Bartlett 124, 172
    conformity of a proportion125
    equality of means 118
    equality of proportions ... 127
    equality of variances......122
    Friedman 172
    of normality.............121
    power...................129
Three-dimensional table 18
Title..........................59
Transposing a matrix 16
Tree
    classification 199
    decision.................199
    regression 199
Type of an object...............7

V

v-test 245, 254
Variance
    estimation 261
    test of equality...........122
Vector 9
    selection 12
Ventilation
    factor.............39, 49, 293
    ordinal factor 294

W

Work session...................3